WORKBOOK TO ACCOMPANY

Automotive Maintenance & Light Repair

Rob Thompson

Australia • Brazil • Japan • Korea • Mexico • Singapore • Spain • United Kingdom • United States

© 2014 Cengage Learning. All Rights Reserved. May not be scanned, copied or duplicated, or posted to a publicly accessible web site, in whole or in part.

Workbook to Accompany Automotive Maintenance & Light Repair
Rob Thompson

Senior Product Development Manager: Larry Main

Senior Content Developer: Matthew Thouin

Editorial Assistant: Leah Costakis

Vice President, Marketing: Jennifer Baker

Marketing Director: Deborah S. Yarnell

Marketing Manager: Erin Brennan

Production Director: Wendy Troeger

Production Manager: Mark Bernard

Senior Content Project Manager: Cheri Plasse

© 2014 Delmar, Cengage Learning

ALL RIGHTS RESERVED. No part of this work covered by the copyright herein may be reproduced, transmitted, stored, or used in any form or by any means graphic, electronic, or mechanical, including but not limited to photocopying, recording, scanning, digitizing, taping, Web distribution, information networks, or information storage and retrieval systems, except as permitted under Section 107 or 108 of the 1976 United States Copyright Act, without the prior written permission of the publisher.

> For product information and technology assistance, contact us at
> **Cengage Learning Customer & Sales Support, 1-800-354-9706**
> For permission to use material from this text or product,
> submit all requests online at **www.cengage.com/permissions.**
> Further permissions questions can be e-mailed to
> **permissionrequest@cengage.com**

Example: Microsoft ® is a registered trademark of the Microsoft Corporation.

ISBN-13: 978-1-1113-0742-4

ISBN-10: 1-1113-0742-3

Delmar
5 Maxwell Drive
Clifton Park, NY 12065-2919
USA

Cengage Learning is a leading provider of customized learning solutions with office locations around the globe, including Singapore, the United Kingdom, Australia, Mexico, Brazil, and Japan. Locate your local office at: **international.cengage.com/region**

Cengage Learning products are represented in Canada by Nelson Education, Ltd.

To learn more about Delmar, visit **www.cengage.com/delmar**

Purchase any of our products at your local college store or at our preferred online store **www.cengagebrain.com**

Notice to the Reader
Publisher does not warrant or guarantee any of the products described herein or perform any independent analysis in connection with any of the product information contained herein. Publisher does not assume, and expressly disclaims, any obligation to obtain and include information other than that provided to it by the manufacturer. The reader is expressly warned to consider and adopt all safety precautions that might be indicated by the activities described herein and to avoid all potential hazards. By following the instructions contained herein, the reader willingly assumes all risks in connection with such instructions. The publisher makes no representations or warranties of any kind, including but not limited to, the warranties of fitness for particular purpose or merchantability, nor are any such representations implied with respect to the material set forth herein, and the publisher takes no responsibility with respect to such material. The publisher shall not be liable for any special, consequential, or exemplary damages resulting, in whole or part, from the readers' use of, or reliance upon, this material.

Printed in the United States of America
2 3 4 5 6 7 14

Contents

	Preface	ix
Chapter 1	**Introduction to the Automotive Industry**	1
	Review Questions	1
	Activities	3
	Lab Worksheet 1-1: Identify Safety and Technology Systems on a Late-Model Vehicle	5
Chapter 2	**Safety**	7
	Review Questions	7
	Activities	12
	Lab Worksheet 2-1: Locate and Inspect Jacks and Jack Stands	15
	Lab Worksheet 2-2: Vehicle Hoist Inspection	17
	Lab Worksheet 2-3: Compressed Air System	19
	Lab Worksheet 2-4: First Aid	21
	Lab Worksheet 2-5: MSDS	23
	Lab Worksheet 2-6: Battery Safety	25
	Lab Worksheet 2-7: Chemicals	27
	Lab Worksheet 2-8: Fire Extinguishers	29
Chapter 3	**Shop Orientation**	31
	Review Questions	31
	Activities	37
	Lab Worksheet 3-1: Use a Floor Jack and Jack Stands to Raise and Support a Vehicle	43
	Lab Worksheet 3-2: Hand Tools	45
	Lab Worksheet 3-3: Fasteners	47
	Lab Worksheet 3-4: Bench Grinder Use	49
	Lab Worksheet 3-5: Battery Charger	51
	Lab Worksheet 3-6: Locating Vehicle Information	53
	Lab Worksheet 3-7: VIN ID	55
Chapter 4	**Basic Technician Skills**	57
	Activities	57
	Lab Worksheet 4-1: Communication Skills	93
	Lab Worksheet 4-2: Using Service Information	95
Chapter 5	**Wheels, Tires, and Wheel Bearings**	97
	Review Questions	97
	Activities	104
	Lab Worksheet 5-1: Tire Identification	107
	Lab Worksheet 5-2: Checking Tire Pressure	109

Lab Worksheet 5-3: Checking Tire Wear Patterns 111
Lab Worksheet 5-4: Tire Rotation 113
Lab Worksheet 5-5: Balance Wheel and Tire Assembly 115
Lab Worksheet 5-6: Reinstall Wheel and Torque Lug Nuts 117
Lab Worksheet 5-7: Inspect Wheel and Tire for Air Loss 119
Lab Worksheet 5-8: Tire Pressure Monitoring System Service 121
Lab Worksheet 5-9: Inspect Wheel Bearing 123

Chapter 6 Suspension System Principles 125
Review Questions 125
Activities 129
Lab Worksheet 6-1: Identify Suspension Types and Components—Short/Long Arm 135
Lab Worksheet 6-2: Identify Suspension Types and Components—MacPherson Strut 137
Lab Worksheet 6-3: Identify Suspension Types and Components—Multilink 139
Lab Worksheet 6-4: Identify Suspension Types and Components—I-Beam 141
Lab Worksheet 6-5: Identify Suspension Types and Components—4WD 143
Lab Worksheet 6-6: Identify Suspension Types and Components—Modified Strut 145
Lab Worksheet 6-7: Identify Suspension Types and Components—Rear Suspension 147

Chapter 7 Suspension System Service 149
Review Questions 149
Activities 155
Lab Worksheet 7-1: Suspension Inspection 159
Lab Worksheet 7-2: Identify Causes of Tire Wear 161
Lab Worksheet 7-3: Shock/Strut Inspection 163
Lab Worksheet 7-4: Ball Joint Inspection 165
Lab Worksheet 7-5: Electronic Suspension Inspection 167

Chapter 8 Steering System Principles 169
Review Questions 169
Activities 175
Lab Worksheet 8-1: Identify Steering Linkage Types and Components 179
Lab Worksheet 8-2: Steering Column Inspection 181
Lab Worksheet 8-3: Inspect Power Steering (Hydraulic Assist) 183
Lab Worksheet 8-4: Inspect Power Steering (Electric) 185

Chapter 9 Steering Service 187
Review Questions 187
Activities 192
Lab Worksheet 9-1: Identify SRS Components 193

	Lab Worksheet 9-2: Disable and Enable the SRS	195
	Lab Worksheet 9-3: Steering System Inspection	197
	Lab Worksheet 9-4: Replace a Power Steering Belt	199
	Lab Worksheet 9-5: Inspect Power Steering Fluid	201
	Lab Worksheet 9-6: Inspect for Power Steering Fluid Loss	203
	Lab Worksheet 9-7: Inspect Electric Power Steering (EPS) Assist System	205
Chapter 10	**Brake System Principles**	**207**
	Review Questions	207
	Activities	213
	Lab Worksheet 10-1: Brake Pedal Leverage	221
	Lab Worksheet 10-2: Hydraulics	223
	Lab Worksheet 10-3: Master Cylinders	225
Chapter 11	**Brake System Service**	**227**
	Review Questions	227
	Activities	232
	Lab Worksheet 11-1: Brake Pedal Inspection	235
	Lab Worksheet 11-2: Brake Light Inspection	237
	Lab Worksheet 11-3: Brake Warning Lamp	239
	Lab Worksheet 11-4: Master Cylinder Inspection	241
	Lab Worksheet 11-5: Brake Bleeding	243
Chapter 12	**Drum Brake System Principles**	**245**
	Review Questions	245
	Activities	248
	Lab Worksheet 12-1: Identify Brake Type and Components	251
Chapter 13	**Drum Brake System Inspection and Service**	**253**
	Review Questions	253
	Activities	260
	Lab Worksheet 13-1: General Drum Brake Inspection	261
	Lab Worksheet 13-2: Drum Inspection	263
	Lab Worksheet 13-3: Wheel Cylinder Inspection	265
	Lab Worksheet 13-4: Parking Brake Inspection	267
	Lab Worksheet 13-5: Electric Parking Brake Inspection	269
Chapter 14	**Disc Brake System Principles**	**271**
	Review Questions	271
	Activities	276
	Lab Worksheet 14-1: Identify Disc Brake Types	279
Chapter 15	**Disc Brake System Inspection and Service**	**281**
	Review Questions	281
	Activities	288

	Lab Worksheet 15-1: Disc Brake Inspection	289
	Lab Worksheet 15-2: Identify Pad Wear	291
	Lab Worksheet 15-3: Measure Brake Rotor Thickness and Parallelism	293
	Lab Worksheet 15-4: Measure Rotor Runout	295
Chapter 16	**Antilock Brakes, Electronic Stability Control, and Power Assist**	**297**
	Review Questions	297
	Activities	302
	Lab Worksheet 16-1: ABS Inspection	305
	Lab Worksheet 16-2: Vehicle Modifications	307
	Lab Worksheet 16-3: WSS Testing	309
	Lab Worksheet 16-4: Identify Brake Assist Types	311
	Lab Worksheet 16-5: Vacuum Booster Operation	313
Chapter 17	**Electrical/Electronic System Principles**	**315**
	Review Questions	315
	Activities	322
Chapter 18	**Basic Electrical/Electronic System Service**	**331**
	Review Questions	331
	Activities	337
	Lab Worksheet 18-1: Using a DMM	343
	Lab Worksheet 18-2: Using a Wiring Diagram	347
	Lab Worksheet 18-3: Relay Testing	349
Chapter 19	**Starting and Charging System Principles**	**351**
	Review Questions	351
	Activities	356
	Lab Worksheet 19-1: Identify Starting and Charging Components	361
Chapter 20	**Starting and Charging System Service**	**363**
	Review Questions	363
	Activities	371
	Lab Worksheet 20-1: Battery Inspection	375
	Lab Worksheet 20-2: Battery Open Circuit Voltage Test	377
	Lab Worksheet 20-3: Conductance Testing a Battery	379
	Lab Worksheet 20-4: Battery Capacity Test	381
	Lab Worksheet 20-5: Three-Minute Charge (Sulfation) Test	383
	Lab Worksheet 20-6: Parasitic Draw Testing	385
	Lab Worksheet 20-7: Starter Circuit Testing	387
	Lab Worksheet 20-8: Generator Drive Belt Inspection	389
	Lab Worksheet 20-9: Generator Output Testing	391
	Lab Worksheet 20-10: Charging System Voltage Drop Testing	393

Chapter 21	**Lighting and Electrical Accessories** **395**
	Review Questions 395
	Activities 403
	Lab Worksheet 21-1: Determine Bulb Applications 405
	Lab Worksheet 21-2: Using a Scan Tool to Test Lighting System Operation 407
	Lab Worksheet 21-3: Inspect Wiper/Washer Operation 409
	Lab worksheet 21-4: Check Module Status 411
Chapter 22	**Engine Performance Principles** **413**
	Review Questions 413
	Activities 420
	Lab Worksheet 22-1: Determine Engine Design and Construction 423
Chapter 23	**Engine Mechanical Testing and Service** **425**
	Review Questions 425
	Lab Worksheet 23-1: Cranking Compression Testing—Conventional Compression Tester 431
	Lab Worksheet 23-2: Running Compression Test—Conventional Compression Tester 433
	Lab Worksheet 23-3: Relative Compression Testing 435
	Lab Worksheet 23-4: Compression Testing—Pressure Transducer 437
	Lab Worksheet 23-5: Cylinder Power Balance Test with a Scan Tool 439
	Lab Worksheet 23-6: Engine Vacuum Testing 441
	Lab Worksheet 23-7: Cylinder Leakage Testing 443
Chapter 24	**Engine Performance Service** **445**
	Review Questions 445
	Lab Worksheet 24-1: Relieve Fuel System Pressure 453
	Lab Worksheet 24-2: Remove, Inspect, and Replace Spark Plugs 455
	Lab Worksheet 24-3: Inspect the PCV System 457
	Lab Worksheet 24-4: On-Board Computer Communication Checks 459
	Lab Worksheet 24-5: OBD II Monitor Status 461
	Lab Worksheet 24-6: Retrieve DTCs and Freeze Frame Data 463
Chapter 25	**Automatic and Manual Transmissions** **465**
	Review Questions 465
	Activities 472
	Lab Worksheet 25-1: Identify Automatic Transmission 473
	Lab Worksheet 25-2: Checking Automatic Transmission Fluid Type, Condition, and Level 475
	Lab Worksheet 25-3: Inspecting a Manual Transmission and Checking Fluid Type and Level 477
	Lab Worksheet 25-4: Inspecting a Differential and Checking Fluid Type and Level 479

Chapter 26   Heating and Air Conditioning 481
Review Questions 481
Activities 486
Lab Worksheet 26-1: Perform a Visual Inspection of the Cooling System 489
Lab Worksheet 26-2: Pressure-Test the Cooling System 491
Lab Worksheet 26-3: Determine Correct Coolants... 493
Lab Worksheet 26-4: Inspect Thermostat and Determine Engine
    Operating Temperature 495
Lab Worksheet 26-5: AC Performance Test 497

Chapter 27   Vehicle Maintenance 499
Review Questions 499
Activities 505
Lab Worksheet 27-1: Maintenance Schedules 507
Lab Worksheet 27-2: Check Tire Pressure 509
Lab Worksheet 27-3: Fluid Inspection 511

Preface

The *Workbook to Accompany Automotive Maintenance & Light Repair* is intended to reinforce comprehension of chapter content from the core textbook, to support critical thinking about the material learned, and finally, to allow students to put their knowledge to practice in the shop. Each Workbook chapter includes a section of Review Questions to allow additional review, reflection, and assessment on the chapter content beyond the chapter-end questions in the textbook. Workbook Activities are designed to further ingrain the material in interesting and meaningful ways, such as part/component identification exercises and in-class experiments (see Chapter 22 Activity on Compression Ratio). Finally, Lab Worksheets offer step-by-step, guided instruction through the types of basic inspection, testing, and maintenance procedures that an entry-level technician is likely to perform on the job.

CHAPTER 1: Introduction to the Automotive Industry

Review Questions

1. Not until the introduction of _____ did major changes start to take place in the automotive industry.

2. There are currently about _____ jobs in the United States directly related to transportation.

3. An _____ technician is expected to be able to perform basic inspections and maintenance services.

4. List five basic services and repairs often performed by entry-level technicians.
   a. _____
   b. _____
   c. _____
   d. _____
   e. _____

5. As important as technical skills are, it is also important to be able to _____ and _____ technical information.

6. Which of the following are likely to be the most experienced technicians in the shop?
   a. A technicians
   b. B technicians
   c. C technicians
   d. None of the above

7. Technicians who work in specialty shops that repair and rebuild engine components are called automotive _____.

8. A _____ technician performs body and paint repairs.

9. A _____ _____ communicates with the customer, is knowledgeable about vehicle systems, has sales skills, and often can perform some basic shop operations.

Chapter 1 Introduction to the Automotive Industry

10. Ford ASSET, Toyota T-Ten, and BMW STEP are manufacturer-sponsored training programs available at _____ schools.

11. Describe the difference between a degree and a diploma program.
_____
_____

12. The _____ _____ _____ _____ _____ is a partner with ASE and certifies training programs.

13. Continuing to learn and acquire new skills is called _____ learning.

14. ASE offers _____ general automotive certifications.
    a. four
    b. six
    c. eight
    d. nine

15. Once a person passes ASE tests A1 through A8, he or she is considered a _____ Automobile Technician.

16. ASE Student Certifications are valid for _____ years.

17. Which of the following has been a factor in modern automobile design?
    a. Emissions
    b. Safety
    c. Fuel economy
    d. All of the above

Activities

1. Using an Occupational Outlook Handbook or the OOH sections of the Bureau of Labor Statistics website, research the following information about automotive careers.

   a. Describe skills necessary to be successful as an automotive technician.
   _____
   _____

   b. List some of the tools a technician is likely to use in his or her job.
   _____
   _____

   c. Describe the automotive work environment.
   _____
   _____

   d. Describe the educational options available to further your education in auto technology.
   _____
   _____

   e. What must a person do to become ASE certified?
   _____
   _____

   f. Describe the job outlook in the automotive repair industry.
   _____
   _____

   g. How many people are employed as automotive technicians in the United States?
   _____

   h. What is the salary range for auto technicians?
   _____
   _____

2. Answer the following questions about your school and the automotive program.

   a. Is the school secondary (high school) or post-secondary? _____

   b. If yours is a secondary school, with what post-secondary school does your program partner for continuing education? _____

   c. How long, in semesters, months, or years does it take to complete the auto tech program?
   _____

Chapter 1 Introduction to the Automotive Industry

   d. What are the ASE/NATEF subject areas covered by the auto program?
   _____
   _____

   e. How many of the students obtain jobs in the auto industry during or after completing the auto technology program each year? _____

3. Using the ASE website, www.ase.com, explore ASE certification to complete the following section.

   a. Provide a brief history of ASE and explain how to become ASE certified.
   _____
   _____
   _____
   _____

   b. What are the requirements to become an ASE Master technician?
   _____
   _____

   c. What are the nine basic automotive certifications?
   _____
   _____
   _____
   _____

   d. Describe the X1 Undercar Specialist certification.
   _____
   _____

   e. How and when are ASE tests offered?
   _____
   _____

Lab Worksheet 1-1

Name _____ Date _____ Instructor _____

Year _____ Make _____ Model _____

Engine _____ VIN _____

Identify Safety and Technology Systems on a Late-Model Vehicle

1. Locate the airbags located in the vehicle. Circle each that is installed.

    Driver's side   Passenger side   Side impact   Curtain

2. List other types of safety technology installed on the vehicle. This includes accessories such as backup cameras, rear collision detection, blind spot detection, and many others.
    _____
    _____

3. List the electronic equipment found on the vehicle. This includes navigation, DVD entertainment, Bluetooth, music integration, and other items.
    _____
    _____

4. List passenger comfort accessories found on the vehicle. This includes power seats, heated/cooled seats, dual-zone climate control, and other items.
    _____
    _____

5. Using the vehicle information and the website www.fueleconomy.gov, locate the estimated fuel mileage for this vehicle. _____

    If the vehicle you are using is five years old or older, try to find either the same make and model or a similar vehicle that is five years newer and compare the fuel mileage ratings.

    How do the two vehicles compare? _____

    If there is a difference, why do you think the difference exists?
    _____
    _____

INSTRUCTOR VERIFICATION:

CHAPTER 2 Safety

Review Questions

1. No aspect of automotive repair and service is more critical than _____.

2. PPE stands for:
    a. Proper protective equipment.
    b. Proper protection and equipment.
    c. Personal protective equipment.
    d. Personnel protective equipment.

3. Shop eyewear may consist of which of the following?
    a. Safety glasses
    b. Full face shields
    c. Welding helmets
    d. All of the above

4. Mechanic's work gloves can protect a technician's hands against all of the following except:
    a. Cuts and scrapes.
    b. Bloodborne pathogens.
    c. Mild burns.
    d. All of the above.

5. Losing your _____ is often a gradual process and may be less noticeable than other types of injuries.

6. Back injuries can cause long-term _____ and can bring an early end to a technician's career.

7. Describe the type of footwear required by your school.
    _____

8. A _____ should be worn when working with brake, clutch, or other airborne dust or chemicals.

Chapter 2 Safety

9. Maintaining good personal _____ means keeping your hair washed and combed as well as taking frequent showers and washing your work clothes.

10. Define *work ethic*, and describe what having a good work ethic means to you. _____
_____
_____

11. Make a list of six safe work habits.
   a. _____
   b. _____
   c. _____
   d. _____
   e. _____
   f. _____

12. What is the purpose of marked safety zones in the shop?
_____
_____

13. Never work under a vehicle supported only on a _____ _____ since failure of the hydraulic cylinder will allow the vehicle to drop, causing injury or death.

14. All lifts have a mechanical _____ _____ that should apply as the lift is raised and automatically engages as the lift is lowered.

15. When moving an engine on an engine hoist, lower the engine close to the floor to lower the center of _____ and reduce the chances of the hoist _____ over.

16. What is meant by dressing a tool?
_____
_____

17. Before using a bench grinder, what three items should be checked?
   a. _____
   b. _____
   c. _____

18. Explain why blow guns should never be used while pointing at yourself or another person?
_____
_____

19. List five safety precautions for working with air tools.
    a. _____
    b. _____
    c. _____
    d. _____
    e. _____

20. Describe how to store a creeper when not in use.
    _____
    _____

21. Describe the typical first-aid procedures for minor cuts and scrapes.
    _____
    _____

22. Explain the differences between first-, second-, and third-degree burns.
    _____
    _____
    _____

23. A _____ _____ can be any substance that can impact public health and/or damage the environment.

24. List the four identifiers that classify hazardous wastes.
    a. _____
    b. _____
    c. _____
    d. _____

25. Explain four types of information that are contained in an MSDS.
    a. _____
    b. _____
    c. _____
    d. _____

26. Explain the purpose of the EPA.
    _____
    _____

27. OSHA is responsible for overseeing the _____ and _____ of workers.

28. When the battery charges, it releases hydrogen, which can, if exposed to a spark or flame, cause an _____.

29. List the correct order in which to disconnect and reconnect battery cables.
   _____
   _____

30. The first step when charging a battery is to ensure that the battery charger is _____.

31. When jump-starting a vehicle, which connection should be made last?
   a. Dead vehicle battery positive
   b. Dead vehicle battery negative
   c. Good vehicle battery negative
   d. Dead vehicle engine ground

32. The high-voltage wiring and connections on a hybrid vehicle are _____ in color for easy recognition.

33. Describe three precautions to observe before attempting to start a vehicle.
   a. _____
   b. _____
   c. _____

34. The exhaust gas that causes illness and death if inhaled in sufficient quantity is:
   a. Carbon dioxide.
   b. Oxygen.
   c. Carbon monoxide.
   d. Oxides of nitrogen.

35. Before moving any vehicle, first perform a _____ _____ to ensure the brake system works properly.

36. The by-products from _____ _____, if allowed to come into contact with your skin, cause skin irritation, rashes, and other health concerns.

37. Many aerosols used in the auto shop are _____ and can easily catch fire if used incorrectly.

38. Waste oil and antifreeze should never be allowed to _____ and must be stored in _____ containers.

39. Brake and clutch friction components may contain _____, a compound that with prolonged exposure, causes lung cancer.

40. A type ABC fire extinguisher contains which of the following?
    a. Carbon dioxide
    b. Water
    c. Dry chemicals
    d. All of the above

INSTRUCTOR VERIFICATION:

Activities

1. Develop a list of personal protective equipment necessary for your class.
   _____
   _____
   _____
   _____

2. Match the correct type of PPE with the list of potential shop hazards.

   Mechanic's gloves   Solvent tank or similar chemicals
   Nitril gloves   Bench grinder
   Chemical gloves   Drum brake service
   Safety glasses   Lifting a cylinder head
   Safety boots   Air hammer
   Ear protection   Blood
   Respirator   Mess on floor
   Back brace   Sharp metal

3. Briefly describe how each of the following shop items presents a danger if used improperly.

   Floor jack _____
   _____

   Vehicle hoist _____
   _____

   Engine hoist _____
   _____

   Bench grinder _____
   _____

   Blow gun _____
   _____

   Creeper _____
   _____

   Battery charger _____
   _____

Impact gun _____
_____

4. On a separate piece of paper, make a rough sketch of your lab and include the following safety items:

First aid kit	Emergency shower	Eyewash station
Fire blanket	Power shut-off	MSDS
Fire exit	Fire alarm	Fire extinguishers

Identify the posted evacuation routes.

INSTRUCTOR VERIFICATION:

Lab Worksheet 2-1

Name _____ Date _____ Instructor _____

Locate and Inspect Jacks and Jack Stands.

Locate the lab's floor jacks and jack stands and note any safety concerns.

    Floor jack caster operation _____

    Hydraulic oil leaks _____

    Lock the handle and test jack operation up and down. _____

    Examine the jack stands and note the condition of the stand, the lifting arm, teeth, and release lever. ____
    _____
    _____
    _____

    Locate the load rating capacity for the floor jack and jack stands. _____

    Describe how to properly use the floor jack and jack stands to safely raise and support a vehicle. _____
    _____
    _____
    _____

INSTRUCTOR VERIFICATION: _____

Lab Worksheet 2-2

Name _____ Date _____ Instructor _____

Hoist manufacturer _____ Lifting capacity _____

Type of hoist: Above ground   In-ground   Symmetrical swing arm

            Asymmetrical swing arm   Drive-on

Vehicle Hoist Inspection

1. Examine the swing arms. Move the arms on their pivots, slide the extensions in and out, and inspect and raise the lifting pads. Note your findings.
   _____
   _____

2. Locate the instructions for using the hoist and summarize how to operate.
   _____
   _____

3. Locate any warning decals and summarize the safety precautions for using the hoist.
   _____
   _____

4. What information is necessary to use the hoist correctly and safely?
   _____
   _____

5. Examine the hoist for any signs of hydraulic oil leaks and note your findings.
   _____
   _____

6. Describe how the lift's mechanical safety mechanism operates and how to disengage the safety to lower the lift. _____
   _____
   _____

INSTRUCTOR VERIFICATION:

Lab Worksheet 2-3

Name _____ Date _____ Instructor _____

Compressed Air System

1. Locate and examine compressed air outlets and hoses. Note your findings.
   _____
   _____

2. What is the shop air pressure? _____

3. Examine the quick-disconnect fittings. Note the condition of the fitting and sleeve.
   _____
   _____

4. When the air is not in use, you should turn the hose _____.

5. Describe how to safely use a blow gun to clean dirt from a part.
   _____
   _____

INSTRUCTOR VERIFICATION:

Lab Worksheet 2-4

Name _____ Date _____ Instructor _____

First Aid

1. Describe the location of the first aid kit.
   _____
   _____

2. What are the basic components of the first aid kit?
   _____
   _____

3. In the event that someone in the lab begins to have a seizure, what are the steps to help that person?
   _____
   _____

4. Locate the emergency shower and eyewash station(s). Describe how to operate each.
   _____
   _____

5. With instructor permission, activate the emergency shower and eyewash station to test their operation. Note your findings.
   _____
   _____

INSTRUCTOR VERIFICATION:

Lab Worksheet 2-5

Name _____ Date _____ Instructor _____

MSDS

1. Describe the location of the MSDS book. _____

2. Why is knowing the location of the MSDS important?
   _____
   _____

3. In addition to the lab MSDS book, where else can MSDS information be found?
   _____
   _____

4. Select three different MSDSs for three items used in the lab and find the following information:
   a. Product name _____ Product type _____
   b. Hazard classification: Flammable   Toxic   Reactive   Corrosive
      If flammable, what is the flash point? _____
      If toxic, by what sort of exposure? _____
      If reactive, with what other agent(s)? _____
      If corrosive, what is the pH? _____
   c. Recommended PPE _____
   d. Handling and disposal requirements _____
   e. Proper first aid _____

   a. Product name _____ Product type _____
   b. Hazard classification: Flammable   Toxic   Reactive   Corrosive
      If flammable, what is the flash point? _____
      If toxic, by what sort of exposure? _____
      If reactive, with what other agent(s)? _____
      If corrosive, what is the pH? _____
   c. Recommended PPE _____
   d. Handling and disposal requirements _____
   e. Proper first aid _____

a. Product name _____ Product type _____

b. Hazard classification: Flammable   Toxic   Reactive   Corrosive

   If flammable, what is the flash point? _____

   If toxic, by what sort of exposure? _____

   If reactive, with what other agent(s)? _____

   If corrosive, what is the pH? _____

c. Recommended PPE _____

d. Handling and disposal requirements _____

e. Proper first aid _____

INSTRUCTOR VERIFICATION:

Lab Worksheet 2-6

Name _____ Date _____ Instructor _____

Battery Safety

1. Explain the hazards associated with automotive batteries.
   _____
   _____

2. By which hazardous material category are batteries classified? _____

3. Locate a battery either in the lab or in a vehicle. Describe battery location.
   _____

4. Is there any evidence of acid leaks and/or corrosion?   Yes _____ No _____

5. Describe how battery acid and corrosion can be neutralized.
   _____
   _____

6. When disconnecting a battery from the vehicle, list the proper order in which to remove and reinstall the battery cable connections.
   _____
   _____

7. When connecting a battery charger to a battery removed from the vehicle, list the steps to properly connect the battery charger.
   _____
   _____

8. What color are the high-voltage wiring and components in a hybrid vehicle?
   _____

INSTRUCTOR VERIFICATION:

Lab Worksheet 2-7

Name _____ Date _____ Instructor _____

Chemicals

1. Compile a list of commonly used chemicals in the auto lab. Include at least five chemicals. _____
   _____
   _____

2. Which of these chemicals have special handling and disposal requirements?
   _____
   _____

3. How and where is waste coolant stored for your lab?
   _____
   _____

4. How and where is waste oil stored for your lab?
   _____
   _____

5. Why are waste coolant and waste oil products stored separately?
   _____
   _____

6. Describe the type of solvent cleaning system used in your lab.
   _____
   _____

7. How is waste solvent handled and disposed of?
   _____
   _____

INSTRUCTOR VERIFICATION:

Lab Worksheet 2-8

Name _____ Date _____ Instructor _____

Fire Extinguishers

1. Note the location and type of each fire extinguisher for your lab.
   _____
   _____
   _____
   _____
   _____
   _____

2. Note any of the extinguishers that need to be recharged based on either their service date or the charge indicator scale. _____

3. List the three most common fire classes.
   _____
   _____
   _____

4. What are the contents of each of the three most common types of fire extinguishers?
   _____
   _____

5. Why are there different extinguisher types for different types of fires?
   _____
   _____

6. What could be the consequences of using the wrong type of extinguisher on a fire?
   _____
   _____

7. Explain how to use a fire extinguisher to put out a small gasoline fire.
   _____
   _____
   _____
   _____

Shop Orientation

Review Questions

1. One of the main goals of the program instructor is to teach you how to perform repairs _____ and by the manufacturer's recommended service procedures.

2. List four common areas or items that are part of the cleanup routine.
   a. _____
   b. _____
   c. _____
   d. _____

3. The primary purposes of a teaching environment are _____ and _____.

4. The primary goals of a repair shop are to repair vehicles quickly and _____ and to make a _____.

5. A new employee often must first complete a _____ period, during which his or her attendance, dependability, and initiative are closely monitored.

6. Explain why as a new employee you are not likely to be the highest-paid technician in the shop?
   _____
   _____
   _____

7. List three reasons to maintain a clean shop.
   a. _____
   b. _____
   c. _____

8. Shop cleanup often starts with making sure your _____ and the shop tools are clean, organized, and _____ to the proper location.

31

9. Which of the following may be part of an entry-level employee's job responsibilities?
    a. Shuttling customers
    b. Shop housekeeping
    c. Obtaining parts
    d. All of the above

10. Describe what is included in performing a PDI.
    _____
    _____

11. The idea that you will need to continue your education and training as a technician is called _____ _____.

12. A tool that is broken or damaged can result in lost _____ for the technician and could even cause personal _____ or damage to a vehicle.

13. Explain the uses of the box end and the open end of a combination wrench.
    _____
    _____

14. When trying to remove a tight bolt, a technician should use which of the following?
    a. Open-end wrench
    b. Socket and ratchet
    c. Pliers
    d. Any of the above

15. Which is not a common type of socket?
    a. 12 point
    b. 8 point
    c. 6 point
    d. Shallow

16. All of the following are common ratchet and socket drive sizes except:
    a. ¼ inch
    b. ⅛ inch
    c. ¾ inch
    d. ½ inch

17. Describe the care and maintenance of a ratchet to keep it in good working condition.
    _____
    _____

18. When discussing screwdriver types: Technician A says Torx and Phillips are similar enough that each can be used in place of the other. Technician B says Phillips screw heads are commonly used as brake fasteners. Who is correct?
    a. Technician A
    b. Technician B
    c. Both A and B
    d. Neither A nor B

19. When installing a hubcap, which type of hammer may be used?
    a. Ball peen
    b. Dead blow
    c. Rubber
    d. Brass

20. What are the three uses for a test light?
    a. _____
    b. _____
    c. _____

21. Air-operated _____ _____ are often used to remove lug nuts and other tight fasteners.

22. When trying to align two holes to install a fastener, a _____ _____ may be used.

23. Click, dial, beam, and digital are all types of _____ _____.

24. Identify the tools shown in Figure 3-1.

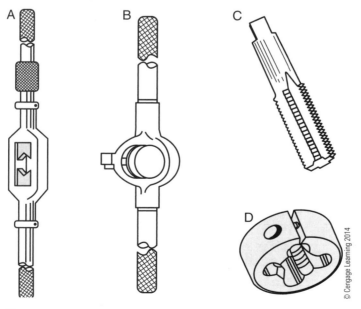

Figure 3-1 Tap and die tools.

a. _____
b. _____
c. _____
d. _____

25. When using a wrench to loosen or tighten a fastener, use the _____ end of the wrench.

26. Never use _____ in place of the correct tool since they can damage fasteners and other parts.

27. When loosening or tightening a fastener, you should pull toward yourself/push away from yourself (circle the correct choice).

28. Explain how to properly set a click-type torque wrench.
_____
_____

29. When a threaded hole is so damaged that the threads cannot be repaired, a _____ _____ can often be used to fix the damage.

30. Tool safety begins with using the _____ tool for the job.

31. Air tools should be _____ daily and checked for proper operation each time they are used.

32. List four examples of tools commonly supplied by the shop.
a. _____
b. _____
c. _____
d. _____

33. Bench _____ are used to reshape metal and dress tools.

34. A _____ _____ can be either petroleum- or water-based and is used to remove dirt, oil, grease, and other substances from parts.

35. List six steps to properly connect and disconnect a battery charger.
a. _____
b. _____
c. _____
d. _____
e. _____
f. _____

36. Shop air pressure should be regulated to:
   a. 60 psi
   b. 90 psi
   c. 120 psi
   d. It is not regulated

37. ________________ and ________________ covers are used to help protect the vehicle during service.

38. Identify the parts of the bolt shown in Figure 3-2.

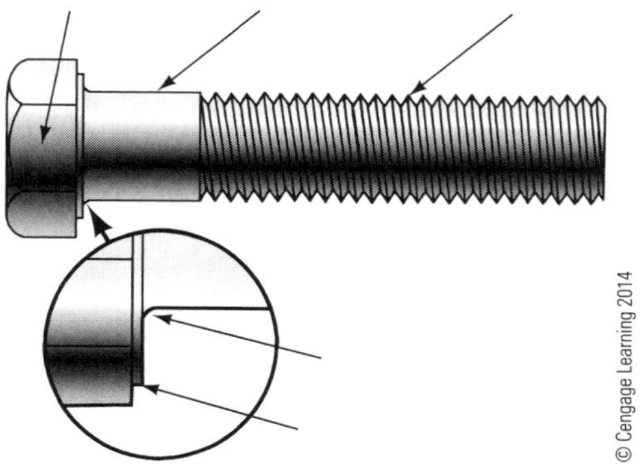

Figure 3-2 Bolt part ID.

39. Describe how to properly measure a bolt to determine its dimensions.

____

____

40. List seven steps to properly maintain a torque wrench.
   a. ____
   b. ____
   c. ____
   d. ____
   e. ____
   f. ____
   g. ____

41. Explain why is it important to always lower the lift onto the safety mechanism before working under the vehicle.

____

____

42. To use a floor jack to raise a vehicle, you must turn the handle ________________.

© 2014 Cengage Learning. All Rights Reserved. May not be scanned, copied or duplicated, or posted to a publicly accessible web site, in whole or in part.

36 Chapter 3 Shop Orientation

43. List the eight steps to lift and support a vehicle with a floor jack and jack stands.

   a. _____
   b. _____
   c. _____
   d. _____
   e. _____
   f. _____
   g. _____
   h. _____

44. Define the following acronyms:

   a. FWD
   b. RWD
   c. 4WD
   d. AWD

45. Describe the vehicle identification number and where it is usually located on a vehicle.

   _____
   _____

46. What type of information is contained on the vehicle emission control information (VECI) decal?

   _____
   _____

47. Which of the following is often contained on a door decal?

   a. Tire pressure
   b. Paint codes
   c. Gross vehicle weight
   d. All of the above

48. What is the purpose of the primary and secondary hood release mechanisms?

   _____
   _____

49. Describe six components of a service order.

   a. _____
   b. _____
   c. _____
   d. _____
   e. _____
   f. _____

Activities

1. Identify the tools shown in Figure 3-3 (a through g).

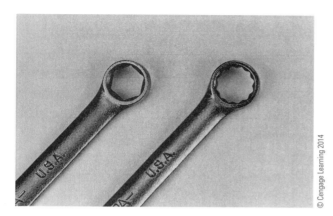

a. _____

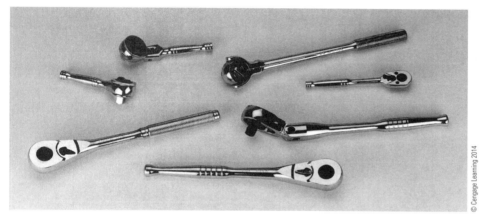

b. _____

c. _____

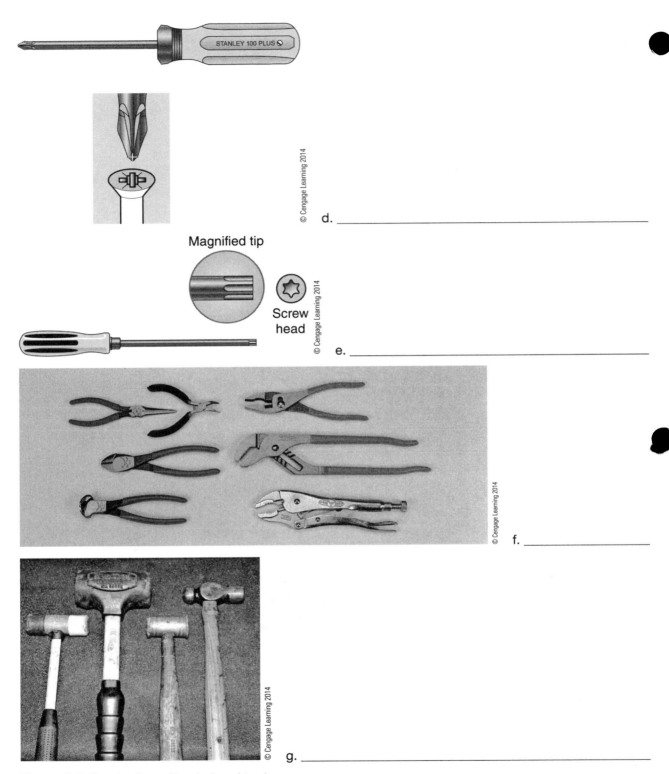

d. _____

e. _____

f. _____

g. _____

Figure 3-3 A selection of basic hand tools.

2. Describe the maintenance required to keep ratchets, screwdrivers, punches, and chisels in good working condition.

   _____

   _____

3. Identify the air tools shown in Figure 3-4.

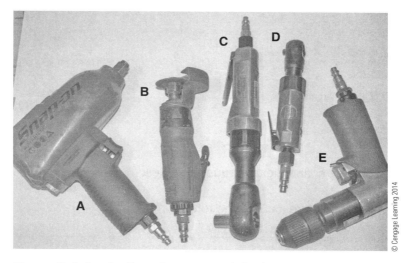

   Figure 3-4 A selection of common air tools.

   a. _____

   b. _____

   c. _____

   d. _____

   e. _____

   Describe the maintenance required to keep air tools in good working condition.

4. Label the parts of the torque wrenches shown in Figure 3-5 (a through d) by writing the labels where indicated on each photo.

   a.

   b.
   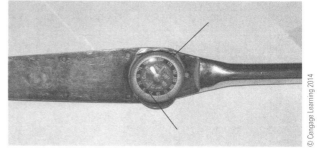

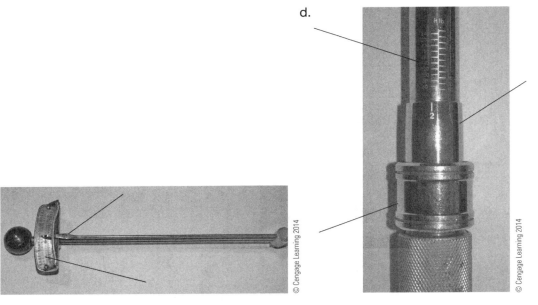

Figure 3-5 Click, beam, and digital torque wrenches.

Explain how to properly set and store a click-type torque wrench.

_____
_____

5. Identify the lifts used in your lab. Describe each lift and how it operates. If there are several lifts of the same manufacturer and type, just describe one of the lifts.

   Lift 1. _____
   _____

   Lift 2. _____
   _____

   Lift 3. _____
   _____

6. Match the following vehicle brands to the manufacturer.

Buick	Chrysler
Cadillac	Ford
GMC	Honda
Chevrolet	Toyota
Scion	BMW
Dodge	General Motors
Ram	
Lexus	
Acura	
Mini	
Pontiac	
Lincoln	

Lab Worksheet 3-1

Name _____ Date _____ Instructor _____

Year _____ Make _____ Model _____

Use a Floor Jack and Jack Stands to Raise and Support a Vehicle

1. Locate and note the recommended lifting and jacking points for this vehicle.
   _____

2. Before lifting, the transmission should be placed in _____.

3. Place a set of wheel chocks against the front and back of a tire on the axle that is not being raised.

4. Position the floor jack under a lift point and raise the vehicle slightly. Have your instructor check before proceeding.
   Instructor Check _____

5. Continue to raise the vehicle until the wheel(s) are off the floor. Position the jack stand(s) under the vehicle where it is to be supported. If using two jack stands, ensure that both are set to the same height. Note the location for the jack stands.

6. Slowly release the floor jack so that the vehicle settles onto the jack stand.
   Instructor Check _____

7. Recheck that the vehicle is safely contacting the jack stand and that the stands are fully supporting the vehicle.
   Instructor Check _____

8. To lower the vehicle, position the floor jack under a lift point. Raise the jack until the vehicle lifts up off the jack stands. Remove the jack stands and slowly lower the vehicle to the ground.
   Instructor Check _____

9. Place the transmission back into _____. Remove the wheel chocks.

10. Explain why it is important to locate the correct lifting points when using a floor jack.
    _____
    _____

11. Why is it important for the jack stands to be set to the same height?
   _____
   _____

12. Explain why you should never work on a vehicle supported only with a floor jack.
   _____
   _____

INSTRUCTOR VERIFICATION:

Lab Worksheet 3-2

Name _____ Date _____ Instructor _____

Hand Tools

1. Using your toolkit, locate the English and metric sizes that are close to matching each other in size.

   ½" = _____ mm   14 mm = _____ "   ⅝" = _____ mm

   10 mm = _____ "   ¾" = _____ mm   8 mm = _____ "

   a. Why is it important to use the correct size socket or wrench on a fastener?
   _____
   _____

   b. Describe how to determine if a fastener is English or metric.
   _____
   _____

2. Using a selection of fasteners supplied by your instructor, match the correct size socket to the head of the bolt.

   Bolt 1 _____   Bolt 2 _____   Bolt 3 _____

   Bolt 4 _____   Bolt 5 _____   Bolt 6 _____

INSTRUCTOR VERIFICATION:

Lab Worksheet 3-3

Name _____ Date _____ Instructor _____

Fasteners

Using a selection of bolts supplied by your instructor, determine the correct thread pitch, length, and diameter for each.

1. TP _____   Diameter _____   Length _____

2. TP _____   Diameter _____   Length _____

3. TP _____   Diameter _____   Length _____

4. TP _____   Diameter _____   Length _____

5. TP _____   Diameter _____   Length _____

6. TP _____   Diameter _____   Length _____

7. What will happen if the incorrect bolt thread is inserted and tightened into a threaded hole?
   _____
   _____

INSTRUCTOR VERIFICATION:

Chapter 3 Shop Orientation 49

Lab Worksheet 3-4

Name _____ Date _____ Instructor _____

Bench Grinder Use

1. Label the parts of the bench grinder shown in Figure 3-6 (a through d).

Figure 3-6 Bench grinder.

a. _____
b. _____
c. _____
d. _____

2. Examine the eye shields, grinder wheels, and power cord before use. Note your findings: _____
_____

3. Turn the grinder on and listen for any noise that may indicate a problem with the grinder. Examples can be growling from the bearings or from loose debris in the wheel covers. Do not use the grinder if there is any damage to the grinder wheels. Note your findings. _____
_____

4. With the grinder off, make sure the work rest is tight and positioned close to the wheel. Turn the grinder on and place a worn chisel on the work rest. The chisel should be placed as shown in Figure 3-7.

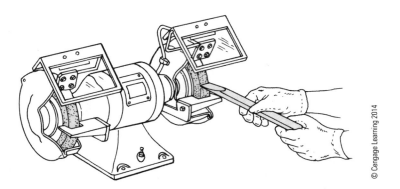

Figure 3-7 Positioning of chisel on bench grinder for sharpening.

5. Move the chisel back and forth across the grinder wheel until properly dressed.

INSTRUCTOR VERIFICATION:

Lab Worksheet 3-5

Name _____ Date _____ Instructor _____

Year _____ Make _____ Model _____

Battery Charger

1. Identify the battery: Brand _____ Group _____

   CCA _____ CA _____ RC _____ AH _____

2. Measure and record battery voltage. _____

3. Using the charging rate information on the battery charger, record the recommended charging rate and time: Rate _____ Time _____

4. Ensure that the charger is off and unplugged from the electrical outlet. Connect the charging cable clamps to the correct posts on the battery.

   Instructor Check _____

5. Plug the charger into an electrical outlet, turn on the charger, and set the charge rate.

   Instructor Check _____

6. Once the charging is complete, turn the charger off and remove the charging cables.

7. After the battery has stabilized for approximately five minutes, measure and record the battery voltage.
   Voltage _____

8. Based on the battery voltage, record your conclusions about the battery. _____
   _____

INSTRUCTOR VERIFICATION:

Lab Worksheet 3-6

Name _____ Date _____ Instructor _____

Year _____ Make _____ Model _____

Locating Vehicle Information

1. Locate and record the VIN: VIN _____

   Location _____

2. Using the decal(s) in the door jamb, locate and record the following information:

   Build date _____   Gross weight _____

   Tire pressure _____   Tire size _____

3. Locate the vehicle emission control identification (VECI) decal and record the following information:

   Emission year _____   Engine size _____

   Tier/Bin _____   Installed emission devices _____

4. For General Motors vehicles, locate the SPO tag and note its location:

   SPO tag location _____

   What is the purpose of the SPO tag? _____

INSTRUCTOR VERIFICATION:

Lab Worksheet 3-7

Name _____ Date _____ Instructor _____

Year _____ Make _____ Model _____

VIN ID

1. Locate and record the VIN: VIN _____

   Location _____

2. What type of information is contained in the VIN? _____
   _____

3. Using service information, determine the following based on the VIN:
   a. Model year _____
   b. Country of manufacture _____
   c. Engine size _____
   d. Manufacturer/division _____
   e. Body/platform type _____
   f. Serial number _____
   g. Check digit _____

4. Examine the vehicle and note other locations where the VIN can be found:
   a. _____
   b. _____
   c. _____
   d. _____

Why is the VIN located in more than one place? _____

INSTRUCTOR VERIFICATION:

Basic Technician Skills

Activities

I. Employability Skills

1. Describe what professionalism means to you.

2. Define your idea of work ethic.

3. How would other people describe your work ethic?

4. Motivation and initiative are important for auto technicians; explain what motivation and initiative mean to you.

5. Define the following terms as an employer's expectations of an employee.

Positive attitude

Sense of responsibility

Initiative

Flexibility

___

Dependability

___

Productivity

___

Enthusiasm

___

6. Develop a list of several personal intrinsic and extrinsic motivational factors.

___

II. Oral Communication

1. Why is it important to have good oral communication skills in the workplace?

___

2. What can result from not being able to communicate clearly and professionally when working with customers?

___

3. Define *slang*.

___

4. List several slang terms you use in everyday conversation and the corresponding standard English meanings for the terms.

___

III. Nonverbal Communication

1. Describe three types of nonverbal forms of communication.

   _____
   _____

2. When dealing with others in a professional environment, what forms of nonverbal communication should you be careful to avoid?

   _____
   _____

IV. Reading and Writing Skills

1. Technicians must be able to read, understand, and apply what they read in order to perform tests and repairs. Following are words commonly used in technical service information with which you need to be familiar. Using a separate piece of paper, define each of the following verbs:

   adjust _____   align _____
   analyze _____   assemble _____
   balance _____   bleed _____
   charge _____   check _____
   clean _____   correct _____
   determine _____   diagnose _____
   disassemble _____   discharge _____
   evacuate _____   flush _____
   hone _____   inspect _____
   locate _____   measure _____
   perform _____   purge _____
   remove _____   replace _____
   resurface _____   service _____
   test _____   torque _____
   verify _____

2. The following paragraph is based on actual service information; read the paragraph and summarize the basic concept.

   "When the ignition is switched ON, the transponder embedded in the key is energized by the exciter coils surrounding the ignition lock cylinder. The energized transponder transmits a unique signal value, which is received by the theft deterrent control module (TDCM). The TDCM compares this signal value to a value stored in read-only memory (ROM) as the learned key code. The TDCM sends a randomly generated number to the transponder, which is called a challenge. Both the transponder and the TDCM perform a calculation on the challenge. If the calculations match, the TDCM sends the fuel enable password via the serial data circuit to the engine control module (ECM). The signal is sent over the controller area network (CAN) high-speed data bus.

If either the transponder's unique value or the calculation to the challenge is incorrect, the TDCM will send the fuel disable password to the ECM via the serial data circuit."

List any words used with which you are unfamiliar:
___

Using a dictionary, locate the words you listed above as unfamiliar and define each in your own words.
___
___
___

Summarize the paragraph in your own words.
___
___
___
___
___

V. Penmanship

Examine the repair order example in Figure 4-1 to answer the following:

Figure 4-1 Poorly written repair order.

1. What is the customer's name? _____

2. Customer address? _____

3. Vehicle year, make, model? _____

4. Customer concern? _____

5. Action required to determine cause of concern? _____
   _____

6. If you were the customer presented with the RO, what would be your reaction?
   _____
   _____

7. Imagine that you are a judge or a member of a jury and were making a decision about a disputed repair based on this RO. Describe your reaction to the RO.
   _____
   _____
   _____
   _____

VI. Spelling

Being able to properly spell, especially automotive terms, is an important skill for communicating with co-workers and customers. In the following example sentences, identify (circle) the misspelled words.

1. Customer states her breaks are making noise.

2. The custemer says the steering wheel is shaking when braking.

3. Rotated tires and torked wheels.

4. Replaced spark plugs, air filter, and fule filter.

5. Performed too wheel alingment, rotated and balanced the tires.

6. Check braks, making noise when stoping.

7. Left front shok leeking.

VII. Grammar

Read the following passage and answer the questions for each section.

(1) Torque converter's transfer engine torque from the engines crankshaft to the transmission, (2) doing the need away for a mechanical clutch too couple the transmission too the engine. (3) Modern torque converters get this power transfer by fluid using the following major component—the cover or shell, impeller, turbine, and stator. (4) All late model torque converter's include a lock-up clutch, to improve vehicle's fuel economy. (5) Good diagnosis of the customers complaints is needed since some complaints can be masked by faults in other powertrain systems.

1. Choose the best correction for section 1 of the passage.
    a. Torque converters transfer engine torque from the engines' crankshaft to the transmission.
    b. Torque converters transfer engine torque from the engine's crankshaft to the transmission.
    c. Torque converters transfer engine torque from the engine crankshaft to the transmission.
    d. Torque converter's transfer engine torque from the engine's crankshaft to the transmission.
    e. No correction necessary.

2. Choose the best correction for section 2 of the passage.
    a. Doing away for a mechanical clutch to couple the transmission to the engine.
    b. This does away with a mechanical clutch to couple the transmission to the engine.
    c. This, does away a mechanical clutch to couple the transmission to the engine.
    d. Doing away with a mechanical clutch, to couple the transmission to the engine.
    e. No correction necessary.

3. Choose the best correction for section 3 of the passage.
    a. Modern torque converters, get this power transfer by using the following major component, the cover or shell, impeller, turbine, and stator.
    b. Modern torque converters, transfer fluid using the following major components: the cover or shell, impeller, turbine, and stator.
    c. Modern torque converters get this power transfer by using the following major components: the cover or shell, impeller, turbine, and stator.
    d. Modern torque converters get this power transfer, using the following major component—the cover or shell, impeller, turbine, and stator.
    e. No correction necessary.

4. Choose the best correction for section 4 of the passage.
    a. All late-model torque converters include a lock-up clutch to improve vehicle fuel economy.
    b. All late-model torque converter's include a lock-up clutch; to improve vehicle's fuel economy.
    c. All late model torque converter's include a lock-up clutch: to improve the vehicles fuel economy.
    d. All late model torque converters include a lock-up clutch, used to improve vehicle fuel economy.
    e. No correction necessary.

5. Choose the best correction for section 5 of the passage.

   a. Good diagnosis of the customer's complaint is needed, since some complaints, can be masked by faults in other powertrain systems.

   b. Proper diagnosis of the customer's complaints are needed, since some complaints can be masked by faults in other powertrain systems.

   c. Proper diagnosis of the customer's complaint is needed since some complaints can be masked by faults in other powertrain systems.

   d. Good diagnosis of the customer's complaint's is needed since some complaints can be masked by faults in other powertrain systems.

   e. No correction necessary.

VIII. Math Skills

A. Fractions

A fraction is a mathematical way to represent a division of a whole into parts. An example of the fraction ¼ can be show n as Figure 4-2. The parts of a fraction are the numerator, the number on top of the line, and the denominator, the number below the line. The numerator represents the number of parts of the whole that are chosen and the denominator defines the total number of parts.

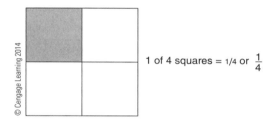

Figure 4-2

Proper fractions like ¼ have a numerator that is smaller than the denominator. When working with SAE fasteners, you will use fractions such as ¼ inch, ⅜ inch, ½ inch, ¾ inch, and similar sizes.

Improper fractions, such as ⅞₃, have the numerator equal to or larger than the denominator. This type of fraction is not often used in automotive applications.

Mixed fractions, such as 2 ½, contain a whole number and a proper fraction. Measurement of bolt lengths, drum brakes, and wheelbase sizes will often use mixed fractions.

Sometimes working with fractions requires you to reduce a fraction. For example, you may measure a bolt that is ¹⁴⁄₁₆ of an inch long. Since both the numerator and denominator are even numbers, both can be reduced to smaller numbers. This is done by factoring. Factoring is used to reduce fractions by listing the prime factors of both the numerator and the denominator. In our example of ¹⁴⁄₁₆, the prime factors of 14 are 1, 2, and 7. The prime factors of 16 are 1, 2, 4, and 8. Since 7 goes into 14 twice and 8 goes into 16 twice, you can reduce ¹⁴⁄₁₆ by writing $\frac{2 \times 7 = 14}{2 \times 8 = 16}$. Since the 2s cancel out, you are left with ⅞.

1. Reduce the following fractions to their smallest form.

   ⁶⁄₈, ¹²⁄₁₆, ¹⁰⁄₃₂, ²⁄₄, ⁸⁄₃₂ _____

2. Place the following fractions in order from smallest to largest.

   ¹³⁄₁₆, ¼, ⅝, ⁹⁄₁₆, ⁷⁄₃₂, ¹¹⁄₁₆, ½, ⁵⁄₁₆, ⁹⁄₃₂, ⅜, ⁷⁄₁₆, ¾, ⅞, ¹⁵⁄₁₆ _____

Fractions can also be converted into decimals, discussed next.

B. Decimals

A decimal is another method of expressing a portion of a whole, similar to a fraction, except unlike a fraction, a decimal does not express the part compared to a number of the whole. Whereas a fraction such as ¼ means that there are four parts in the whole, a decimal is used to show the parts in tenths, hundredths, thousands, or more. This is based on the place value of the number. Place value determines a number's value on either side of a decimal point. A whole number, such as 327, is represented by the ones column, tens column, and one hundreds column, with 3 being in the hundreds, 2 in the tens, and 7 in the ones, as shown in Figure 4-3.

Figure 4-3

To show a portion of the whole number, decimals use numbers to the right of a decimal point, shown in Figure 4-4. The first number to the right of the decimal point has the value of tenths, the second number is hundredths, the third is thousands, and so on. Common measurements in automotive applications use the hundredths and thousandths place, and occasionally even the ten thousandths place.

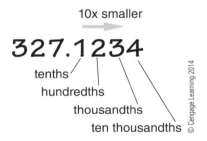

Figure 4-4

To convert a fraction into a decimal, simply divide the numerator by the denominator. For example, to convert ¼ into decimal, divide 1 by 4. If using a calculator, input 1 ÷ 4. Solve the following problems.

1. Find the sum of 21.45, 13.08, and 91.16. _____

2. Subtract 37.18 from 119.05. _____

C. Percentages

A percentage is another method to represent a portion of the whole and is also used to indicate numbers greater than the whole. Percentages are either shown as a number and percent symbol, such as 50%, or written out as 50 percent. This example, 50 percent, means 50 out of 100. Percentages are useful when working with money, such as figuring discounts and markup.

Solving some percentage problems, such as 10 percent off a sale price, is easy. If an item is on sale for $89.95, simply move the decimal point one place to the left to find the price after the discount. In this example, $89.95 is discounted $8.99, or 1/10 of the original price. Two more ways to solve this can be done with a calculator. If the calculator has a % key, you can enter 89.95 and multiply times 10% to find the discounted amount. You can also multiply 89.95 times 90% to find the final price after the discount. If the calculator does not have a % key, you can still find the percentage by multiplying 89.95 by .10 to find the discount amount or by .90

to find the total price after the discount. By using .10 or .90, you are multiplying by the value of the percentage. By using .10, this is equivalent to 1/10 or 10 percent of 1.

Percentages are also used to calculate markup. Markup is the difference between the sale price and what the item actually cost. For example, if you buy a drink from a vending machine for $1 but the cost of the drink to the vendor is only $.25, then the markup on the drink is 400 percent since the drink sells for four times its cost.

Solve the following problems.

1. A part retails for $128.67, but the shop receives a 20 percent discount. What is the cost of the part to the shop? _____

2. A part that cost $47.90 is resold with a markup of 40 percent. At what price is the part sold? _____

3. Sales tax of 6.75 percent is added to all parts and labor. Find the sales tax for the following repair amounts: $19.95, $79.88, $212.45, $420.16. _____

D. Ratios

A ratio provides a mathematical relationship between two items and is written as item A in relation to item B, or A:B. Common examples include gear ratios, rocker arm ratios, braking ratios, and such proportions as mixing water and coolant in a 50:50 ratio. Gear ratios are written as 2:1 and 3.5:1, where the first number represents the number of turns of the driving gear and the second number is the turns of the driven gear, as shown in Figure 4-5.

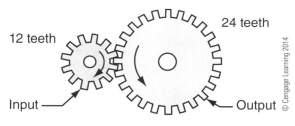

Figure 4-5

Solve the following problems.

Determine the gear ratio based on the following driving to driven gear.

1. Driving gear = 10 teeth, driven gear = 15 teeth _____

2. Driving gear = 20 teeth, driven gear = 50 teeth _____

3. Driving gear = 18 teeth, driven gear = 12 teeth _____

E. Area and Volume

Area is a two-dimensional measurement of an object's length and width. Surface area is calculated by multiplying the length and width of an object. While it is not often you will need to find the area in automotive applications, it is a measurement with which you should be familiar. One example of using area is finding the contact patch size of the tire on the ground. Obtain a contact patch imprint by placing a piece of paper under a tire and allowing the vehicle weight to settle on the tire. Next, remove the paper and measure the size of the imprint of the tire. Calculate the contact patch area and multiply your answer by the recommended tire inflation pressure. Compare your answer to the maximum tire load information located on the tire sidewall.

1. What is the size of the tire's contact patch? _____

2. What is the tire's recommended inflation pressure? _____

3. What is the maximum load rating of the tire? _____

4. How do tire size and pressure affect its load-carrying capacity? _____

Volume is the three-dimensional measurement of length, width, and height. Volume measurement is used when discussing engine size, specifically the engine displacement. This number refers to the volume of air that all of the cylinders can hold. To determine the volume of a cylinder use: cylinder volume = pi/4 × bore^2 × stroke. A slightly simpler method is to use .785 × bore^2 × stroke since .785 is the rough equivalent of pi/4.

If an engine has a bore of 4 inches and a stroke of 3.48 inches, then the cylinder volume is figured: 0.785 × 4^2 × 3.48 = 43.708 cubic inches. Multiply this by the number of cylinders, which is 8, and you get 43.708 × 8 = 349.67, which is close to the engine displacement of the old small block Chevy 350 engine. The same approach can be used for engine sizes in metric. For example, a four-cylinder engine with a bore of 87 mm (8.7 cm) and a stroke of 99 mm (9.9 cm): .785 × 87^2 × 99 = 2,352 cubic centimeters or, as the engine is marketed, as a 2.4L.

Given the formula above, calculate the engine displacements for the following:

1. 8-cylinder, bore 4 inches and stroke 3.5 inches _____

2. 6-cylinder, bore 89 mm and stroke 93 mm _____

3. 4-cylinder, bore 81 mm and stroke 87.3 mm _____

F. English Measurement

English measurement uses the mile, yard, foot, inch, and fractions of the inch. This system has been in use for centuries but has been replaced in most parts of the world with the metric system. However, common uses for English measurement still occur, and you should be able to accurately read an English ruler and tape measure.

1. In Figure 4-6, measure from the starting line (A) to each lettered line and record the measurement in lowest terms. Note: Your measurement should reflect what each lettered line represents on an English ruler and not the actual distance from line A.

2. What is the smallest unit shown in Figure 4-6? _____

In the spaces provided, answer with the correct measurement in lowest terms forthe distances in the figure below.

Example line = 4/16" or 1/4"

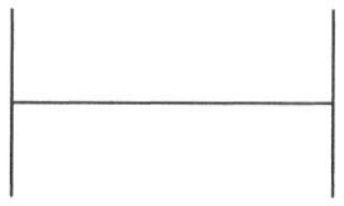

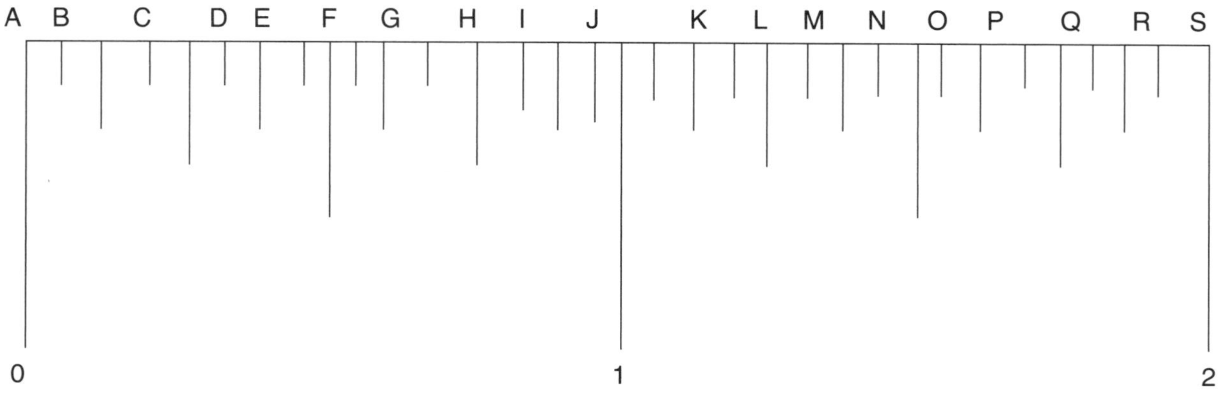
A B C D E F G H I J K L M N O P Q R S
0
1
2

AB_____ 2. AC_____ 3. AD_____ 4. AE_____ 5. AF_____ 6. AG_____

AH_____ 8. AI_____ 9. AJ_____ 10. AK_____ 11. AL_____ 12. AM_____

AN_____ 14. AO_____ 15. AP_____ 11. AQ_____ 17. AR_____ 18. AS_____

© Cengage Learning 2014

Figure 4-6

Using a tape measure, locate the following:

3. Measure and record the wheelbase of a vehicle in the lab. ________________________

4. Measure and record the distance between the two upright columns of an above-ground vehicle lift. __________

5. Measure the width of the lab garage door. ________________________

6. Measure the length of a wiper blade on a lab vehicle. ________________________

Using a ruler, measure and place a mark along the line at the given distance.

7. (1 inch) ________________________ 8. (⅝ inch) ________________________

9. (¾ inch) ________________________ 10. (1³⁄₁₆ inch) ________________________

© 2014 Cengage Learning. All Rights Reserved. May not be scanned, copied or duplicated, or posted to a publicly accessible web site, in whole or in part.

Using an English ruler, measure along the lines in the upper part of Figure 4-7 and record the actual distances in the spaces provided at the bottom of the figure.

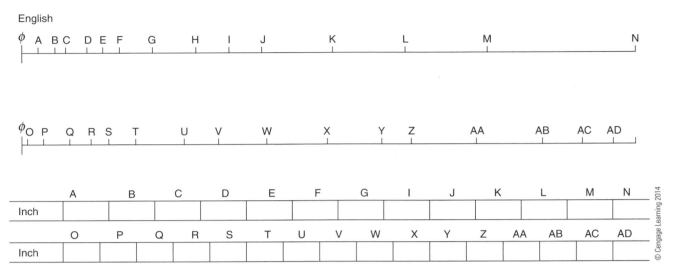

Figure 4-7

G. Metric Measurement

Metric measurement is based on units of 10, which makes switching units very quick and easy since only the decimal place needs to be changed. The metric system uses the meter as its base unit. The meter is then divided into tenths, called decimeters. The decimeter is divided into tenths also, called centimeters. Centimeters are also divided into tenths, called millimeters. There are 10 decimeters, 100 centimeters, and 1,000 millimeters in a meter. 1,000 meters is a kilometer, which is roughly equivalent to 0.62 miles. A meter is roughly 39 inches long, making it slightly longer than a yard.

In Figure 4-8, use a metric ruler to measure from the starting line to each lettered line and record your readings in the spaces provided at the bottom of the figure.

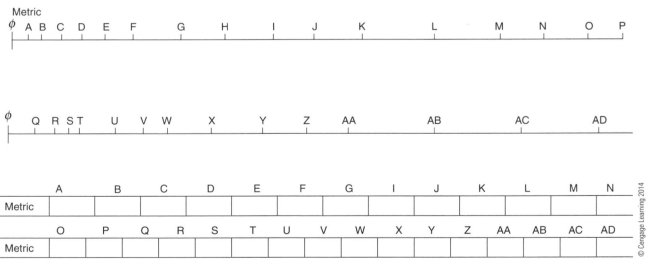

Figure 4-8

Using a metric ruler, locate the following:

1. Length of a bolt provided by your instructor _____

2. Diameter of a fire extinguisher _____

3. Width of a lug nut provided by your instructor _____

4. Length of a wiper blade _____

5. (Instructor option) _____

Using a metric ruler, measure and place a mark along the line at the given distance from the left side of the line.

6. (10 mm) _____   7. (19 mm) _____

8. (13 mm) _____   9. (21 mm) _____

10. (17 mm) _____   11. (14 mm) _____

12. (22 mm) _____   13. (8 mm) _____

14. (12 mm) _____   15. (15 mm) _____

H. Using Measuring in Lab

Using a ruler and/or tape measure, obtain the following measurements:

1. Vehicle _____   Wheelbase _____

2. Vehicle _____   Wiper blades _____

3. Vehicle _____   Lug pattern _____

4. Vehicle _____   Generator pulley diameter _____

5. Vehicle _____   P/S pump pulley diameter _____

I. Labor Time

When a technician performs a job, he or she may be paid based on the labor time to perform the particular job. Labor times are found in estimating guides, also called flat-rate guides. These guides are used to provide an estimate of the amount of time required to perform tasks. For example, the labor time to replace a set of front brake pads may be one hour. If the technician replaced the pads in less than one hour, he or she could then move on to the next job while having been paid for an hour of work even though the actual work time was less. If, however, the technician takes one and a half hours to replace the brake pads, he or she was still be only paid for one hour of work, and so lost money on the repair.

© 2014 Cengage Learning. All Rights Reserved. May not be scanned, copied or duplicated, or posted to a publicly accessible web site, in whole or in part.

Labor guides break an hour down into tenths of an hour, or six minutes. Repair times are given in tenths of hours and are shown in decimal form. For example, a job that pays one half of an hour, labor is shown as 0.5 hours. Jobs may be anywhere from 0.1 hours to several hours depending on the difficulty of the work.

Using an estimating guide, find the labor times for the following items:

1. 2005 Chevy Malibu V6 front brake pad replacement _____

_____

_____

2. 2006 Honda Civic EX sedan timing belt replacement _____

3. 2004 Ford F150 2WD lower ball joint replacement _____

4. 2005 Dodge Durango V8 spark plug replacement _____

5. 2008 Toyota Camry V6 front window motor replacement _____

J. Replacement Parts

Part of servicing and repairing cars and trucks is determining what parts need to be replaced, finding the parts cost and their availability, and preparing the estimate for the customer. Shops purchase parts at a price discount, often called wholesale price, and then mark the parts up and resell them to the customer. This is necessary for the shop to make money and enough profit to stay in business.

When preparing an estimate, you will often need to contact a local part supplier, which may be an aftermarket store like NAPA, O'Reilly's, AutoZone, or similar store, or the parts may be purchased from a local new car dealership. Where the part is purchased depends on the type of part, the cost of the part, the availability, and the quality of the part. In some cases, the correct replacement part can only be bought from the dealer.

Using a school vehicle, your own vehicle, or one provided by your instructor, locate the following part information.

Locating Parts

Select a vehicle, either a school or your own (not a fantasy vehicle) for which to locate parts.

Year _____ Make _____ Model _____

Engine: _____ AT   MT   P/S   A/C   (circle all that apply)

You will need to access either OnDemand5 or AllData for the recommended fluids and then use the following websites, www.napaonline.com, www.autozone.com, or www.advanceautoparts.com to locate the following parts.

1. Engine oil filter Brand _____ Part number _____ Price _____.

2. Engine air filter Brand _____ Part number _____ Price _____.

3. Recommended engine oil weight _____ Brand _____ Price _____.

4. Recommended coolant type _____ Brand _____ Price _____.

5. Wiper blade sizes _____ Brand _____ Price _____.

6. Recommended transmission fluid _____ Brand _____ Price _____.

7. Engine drive belt Brand _____ Part number _____ Price _____.

8. Spark plugs Brand _____ Part number _____ Price _____.

9. Battery Brand _____ Part number _____ Price _____.

Calculate the parts cost to perform an oil change, replace the oil filter, air filter, replace the wiper blades, drive belt, and spark plugs.

10. Total parts cost _____

K. Markup

Markup is the term used to describe the difference between the price paid for a part and the price at which the part is resold to the customer. For example, if the shop buys brake pads for $19 and sells to the customer for $38, the shop marked up the pads 200 percent, selling the pads for twice what the shop paid for them. The amount of markup on a part often varies based on the cost of the part and the manufacturer's suggested price for the part. Parts markup is necessary for the shop to be able to provide a warranty on the parts and labor. If the pads are noisy and the customer returns to have them replaced, the shop usually does not receive reimbursement from the part supplier to replace faulty parts. Therefore the shop performs the work for free.

Contact a local parts store used by your shop and obtain the shop's cost and the list price for the following parts. Then, determine the markup for the parts for a late-model vehicle.

Year _____ Make _____ Model _____

1. Front brake pads   Cost $ _____   List $ _____   Markup $ _____

2. Water pump   Cost $ _____   List $ _____   Markup $ _____

3. Front shock/strut   Cost $ _____   List $ _____   Markup $ _____

4. Air filter   Cost $ _____   List $ _____   Markup $ _____

5. Starter   Cost $ _____   List $ _____   Markup $ _____

Next, determine the resale price of the following parts based on their cost and markup percentages given.

6. Timing belt: cost $29.78, markup 65 percent   Resale price $ _____

7. Radiator: cost $129.13, markup 35 percent.   Resale price $ _____

8. Brake pads: cost $33.49, markup 45 percent   Resale price $ _____

9. Lug nut: cost $1.25, markup 80 percent   Resale price $ _____

10. Upper control arm: cost $103.79, markup 55 percent Resale price $ _____

L. Shop Supplies

Shop supplies are items such as shop rags, chemicals, and small fasteners or hardware that are used during most types of services. Some shops charge a separate fee for these items, often a percentage of the final repair cost up to a set limit. For example, many shops charge a 5 percent shop supply fee on each repair order up to a maximum of $5.

1. Make a list of items used in your shop that you think may be billed as shop supplies.
   _____
   _____

2. Based on the example of 5 percent up to $5, calculate the shop supplies charge for the following:

   a. Total repair bill $123.55 _____
   b. Total repair bill $75.89 _____
   c. Total repair bill $468.33 _____

M. Sales Tax

Sales tax is common in most states and localities and can vary from city to city and county to county. Sales tax can be applied to the total of parts and labor or only to parts or labor, depending on location. When writing estimates and completing repair orders, it is very important for you to correctly determine the sales tax so the customer is charged correctly.

Calculate the sales tax on the following:

1. Total repair bill $378.85, sales tax 6.75 percent _____

2. Total repair bill $39.90, sales tax 7.25 percent _____

3. Total repair bill $234.18, sales tax 5.75 percent _____

4. Determine the sales tax for your location. Tax rate _____

Based on your local sales tax, determine the amount of tax collected for the following:

5. Bill amount $88.89   Sales tax $ _____   Total $ _____

6. Bill amount $246.52   Sales tax $ _____   Total $ _____

7. Bill amount $512.13   Sales tax $ _____   Total $ _____

N. Balancing a Checking Account

Sooner or later you will have a checking or savings account. To keep track of how much money you have, you should periodically balance the account. This means you confirm how much you have in the account by comparing what the bank statement shows to what your own records show. This will help you track your income and expenses and hopefully keep you from overspending your account, which leads to fees and penalties.

To track your money, use a checking account register or a simple running record, like that shown in Figure 4-9. On the left side of the register, record the date and check number for transactions. In the center column note what the transaction was, such as a deposit, cash withdrawal, debit, or similar. In the right columns note the amount of the transaction and in the far right keep a running balance, as shown in Figure 4-9.

AD-Automatic Deposit • AP-Automatic Payment • ATM-Cash Withdrawal • DC-Debit Card • FT-Funds Transfer • SC-Service Charge • TD-Tax Deductible

NUMBER OR CODE	DATE	TRANSACTION DESCRIPTION	PAYMENT, FEE, WITHDRAWAL(−)	✓	DEPOSIT, CREDIT (+)	$ BALANCE
	2/15	Paycheck deposit		✓	487.15	562.77
	2/16	Cash	50	✓		512.77
	2-18	Callahan Auto Parts	78.63	✓		434.14
	2-20	Car payment	288.60	✓		145.54
	2-20	gas	18	✓		127.54
	2/25	insurance	112.25			15.29
	3/1	deposit			390.41	405.70

Figure 4-9

When you receive your account statement, check the balance on the statement and compare it to the balance you show in your register. These will likely not be the same as you may have made many purchases and/or deposits since the statement was printed. To reconcile your balance with that shown in the bank statement, follow these steps:

1. Begin by checking off all transactions that have cleared the bank. This means on your register, check off all debits and deposits that show on the bank statement.

2. Add any uncleared debits to your account to your register balance. If you do not have any uncleared deposits, then your register balance plus uncleared debits should equal the balance shown on the statement.

3. If you have uncleared deposits, subtract the amount from the register balance. If the two numbers do not match, recheck that you have all transactions accounted for in your register and all cleared transactions on the bank statement.

4. If you still cannot find a reason for the numbers not to match, you may have a mistake in your register. Recheck all of your addition and subtraction to check for mistakes.

If your balance and the balance shown by the bank still do not match, go back over every transaction during the statement dates. Make sure you recorded the correct amounts for both deposits and debits in your register. Be sure to add in any interest and subtract any fees, such as ATM fees or account maintenance charges.

IX. Science Skills

A. Energy Conversion

Many forms of energy are used in the automobile, and various energy conversions also take place. As discussed in Chapter 4, energy is not created or destroyed, it is used to perform some type of action. Following are types of energy conversions used in cars and trucks; match the energy type with its function.

Electrical to mechanical   Brake friction

Mechanical to electrical   Combustion

Chemical to thermal   Starter motor operation

Kinetic to heat   Generator operation

Thermal to mechanical

B. Newton's Laws of Motion

Match the following examples with Newton's Laws of Motion.

Newton's First Law – Inertia

Newton's Second Law – Force = mass · acceleration

Newton's Third Law – Equal and opposite reaction

Weight transfer during braking _____

Accelerating from a stop _____

Brake caliper operation _____

Sliding to the outside of a turn while cornering _____

Increasing the weight of the vehicle but leaving the power output unchanged _____

C. Forces

To slow or stop the vehicle, all you have to do is press the brake pedal. But do you understand what takes place each time that pedal is pressed? The pedal itself is an application of one of the oldest forms of machines, the lever. Known since ancient times, the lever is not only one of the oldest machines, it is also one of the most simple. A lever applies force to gain mechanical advantage—for instance, when moving a heavy object. This is accomplished by increasing the distance over which the force is applied.

A teeter-totter is another example of a lever. If two people of the same weight sit the same distance apart from the middle, the teeter-totter will balance and both people will be suspended off the ground at equal heights. But what happens if one person weighs more than the other? To offset the imbalance, the heavier person must move forward, closer to the middle, or the lighter person must move farther away from the middle, to the very edge of the board, as shown in Figure 4-10.

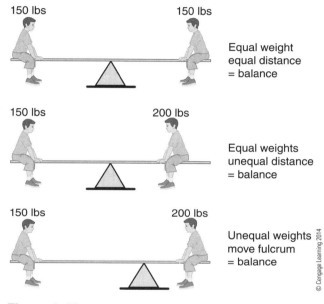

Figure 4-10

1. If the fulcrum stays in the middle, why must one person move to offset the imbalance?

_____

2. If the two people stay in place, how must the fulcrum move to rebalance the teeter-totter?

_____

A teeter-totter is an example of a first-class lever. In a first-class lever, the fulcrum is located between the load and the effort, as shown in Figure 4-11. Using a shovel to dig up a plant is another example of a first-class lever in action.

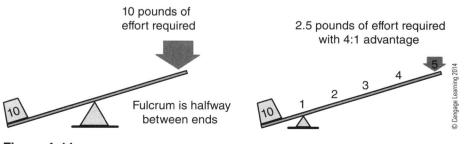

Figure 4-11

The brake pedal in a car operates as a second-class lever. With a second-class lever, the fulcrum is at one end of the lever instead of the middle. The effort is applied to the other end of the lever, and the force is applied somewhere in between the effort and the fulcrum, as shown in Figure 4-12. If you look under the dashboard of a vehicle, you will be able to see how the brake pedal is mounted. Notice that the top of the pedal is the mounting point, which acts as the fulcrum. Below the fulcrum the pushrod is attached, which extends forward through the firewall to the brake booster and/or master cylinder. The footpad at the bottom of the pedal is where the effort is applied.

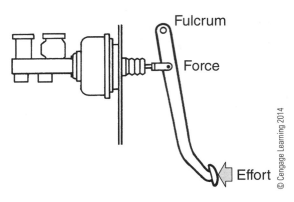

Figure 4-12

3. Measure the length of the pedal between the pushrod and fulcrum and between the pushrod and the lowest point of the pedal for several vehicles.

   Pushrod-to-fulcrum length _____   Pushrod-to-pedal length _____

4. Calculate the ratio of the lever by dividing the distance from the pushrod to the pedal by the distance from the pushrod to the pivot or fulcrum.

   Pushrod to pedal _____ /pushrod to pivot _____ = _____

   This ratio represents how much mechanical advantage is gained by the leverage of the brake pedal.

5. If the driver applies 50 pounds of force to the brake pedal, how much force is delivered by the brake pushrod? _____

6. If the driver applies 75 pounds of force to the brake pedal, how much force is delivered by the brake pushrod? _____

7. If the driver applies 150 pounds of force to the brake pedal, how much force is delivered by the brake pushrod? _____

8. In Figure 4-13, label the lever, pivot, and where the force is being applied.

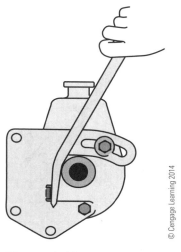

Figure 4-13

A gear is a circular component that transmits rotational force to another gear or component. Gears have teeth so that they can mesh with other gears without slipping. The amount of leverage or mechanical advantage gained by the gear and ultimately, the gear ratio, is determined by the distance from the center of the gear to the end of a tooth, as shown in Figure 4-14.

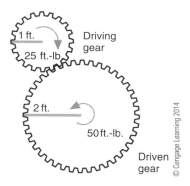

Figure 4-14

In Figure 4-14, the driving gear has a radius of 1 foot. As the driving gear turns clockwise, the driven gear rotates counterclockwise. The driven gear can apply twice the torque of the driving gear since the driven gear has a radius of 2 feet. By using a gear that is larger, more force can be applied since the distance from the center of the gear to the teeth creates leverage. This advantage allows gears to transmit a large amount of torque through the drivetrain to propel the vehicle.

Since the driven gear has twice as many teeth as the driving gear, it will turn at one-half the speed of the driving gear. The relationship between gear radius, and consequently, the number of gear teeth, is called gear ratio. The driving gear will turn twice for every one rotation of the driven gear, so the gear ratio is 2:1, shown in Figure 4-15. If three gears are used, as shown in Figure 4-16, the center gear is an idler gear and does not affect gear ratio. The driving gear and driven gear teeth determine the gear ratio.

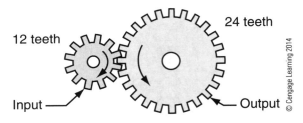

Figure 4-15

Figure 4-16

9. Determine the gear ratios of the gears shown in Figure 4-17. Write your answers in the spaces provided in the figure.

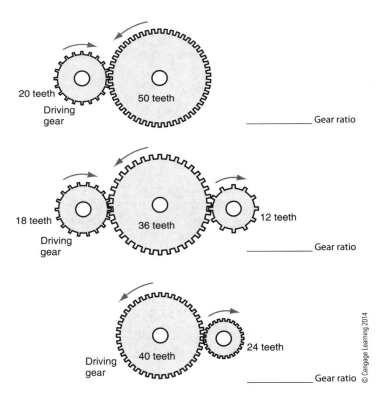

Figure 4-17

Pulleys, along with drive belts, are used to transmit motion. Located on the front of the engine, the crankshaft pulley is used to provide rotational force to other pulleys to drive accessories such as the generator, power steering pump, and water pump. An example of a front-end accessory drive arrangement is shown in Figure 4-18. You may notice that not all of the pulleys are the same size. Using pulleys of different diameters allows the various accessories to be driven at different speeds. For example, generator drive pulleys are typically about half the diameter of the crankshaft pulley. This means that for any given crankshaft rpm, the generator will be spinning about twice as fast.

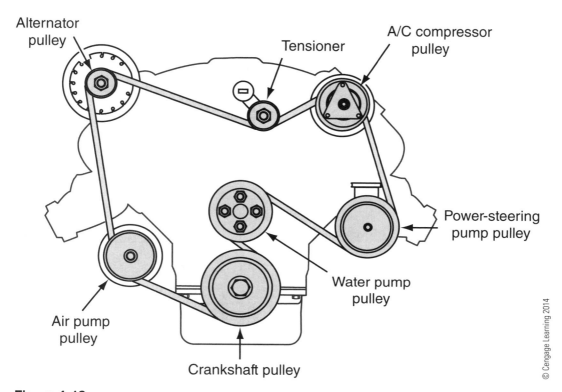

Figure 4-18

Because of the use of multi-rib serpentine belts in modern vehicles, either the rib or the flat side of the belt may drive pulleys. Additionally, some pulleys are driven opposite the engine. Most all newer engines rotate clockwise as seen from the front of the engine. However, some pulleys are driven counterclockwise based on how the drive belt is routed.

10. Examine the drive belt arrangements in Figure 4-19, and label the direction of rotation for each pulley.

80 Chapter 4 Basic Technician Skills

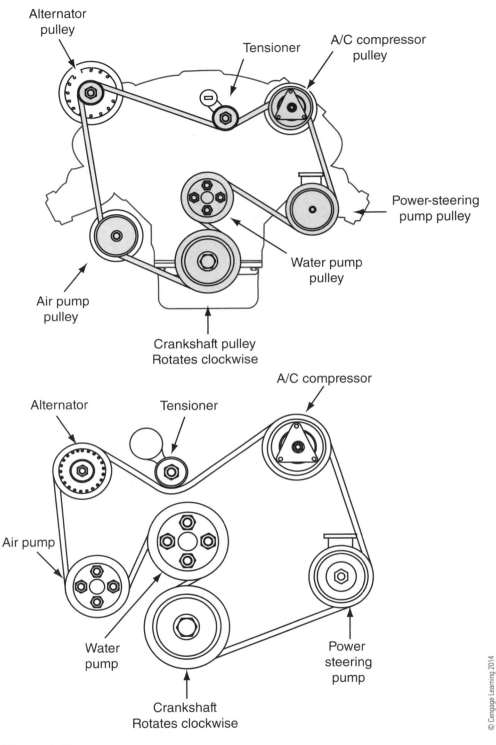

Figure 4-19

D. Friction

In the automotive brake system, the pads and shoes create friction when applied; this in turn creates a lot of heat. Heat is a natural by-product of friction. Rub your hands together briskly and note how fast your skin warms.

1. Why does rubbing your hands together create heat? _____

2. Wet your hands with some water and rub them together briskly again. How did the water change the result? _____

In automotive applications, there is friction used by the brake system, the friction of moving fluids such as motor oil and transmission fluid, friction between the vehicle and the air in which it is traveling, called drag, and friction between the tires and the ground.

Friction is created when uneven surfaces, in contact with each other, start to move relative to each other. Even though the two surfaces may appear smooth, there are slight imperfections that resist moving against each other. The amount of force required to move one object along another is called the coefficient of friction (CoF). Simply put, the CoF is equal to the ratio of force to move an object divided by the weight of the object, $CoF = F/M$. While several factors affect the CoF, such as temperature and speed, we will use a simple example of two bodies moving against each other. Imagine you have a 100 lb (45 kg) block of rubber on the floor of your lab. The force required to slide the block of rubber over concrete would be great. If it takes 100 lb. (45 kg) of force to slide the block, the CoF will be 100/100 or 1.0. Imagine the same block of rubber now sitting on the floor of an ice hockey rink. If it only takes 25 lb (11kg) of force to slide the block, what will the CoF be?

3. 25/100 = _____

To experiment with the CoF of various objects in your lab, you can make a small force gauge using an ordinary ballpoint pen, a rubber band, tape, and a paper clip. Assemble the parts as shown in Figure 4-20. Attach the paper clip to an object and try to pull it across a flat surface using the opposite end of your force meter. Measure the amount of extension of the pen tube from the body of the pen with a ruler. This will give you an idea of how much force is required to drag each object. Record your results below:

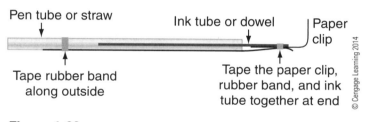

Figure 4-20

4. Object _____ Surface _____ Length of extension _____

5. Object _____ Surface _____ Length of extension _____

6. Object _____ Surface _____ Length of extension _____

7. Why did some objects require more force than others did? _____

8. What effect does the surface used to slide across affect the CoF? _____

9. If a liquid is placed between the two objects, what effect will that have on the CoF? _____

10. What could happen if the CoF of the brake pads or shoes was too high? _____

11. What could happen if the CoF of the brake pads or shoes was too low? _____

12. What factors do you think are involved in determining the correct CoF for a particular vehicle? _____

82 Chapter 4 Basic Technician Skills

The CoF of brake linings can have an impact on how well a vehicle stops. Using lining materials with too high or low a CoF can cause brake performance issues, rapid wear, and customer dissatisfaction. Table 4-1 shows common brake material coefficients.

DOT Edge Code	Coefficient of Friction @ 250° F and 600° F	Fade Probability
CC	0.0 to 0.15 both temps	
DD	0.15 to 0.25 both temps	
EE	0.25 to 0.35 both temps	0-25% @ 600° F
FE	0.25 to 0.35 @ 250° F temp 0.35 to 0.45 @ 600° F	2% to 44% fade at 600° F
FF	0.35 to 0.45 @ both temps	0-22% fade at 600° F
GG	0.45 to 0.55	Very rare
HH	0.55 to 0.65	Carbon/Carbon only - glows at about 3000° F
Organic linings	cold: 0.44, warm: 0.48	
Semimetallic	cold: 0.38, warm 0.40	
Metallic	cold: 0.25, warm: 0.35	
Synthetic	cold: 0.38, warm: 0.45	

E. Hydraulics

Hydraulics refers to using fluids to perform work. For practical purposes, fluids can be pressurized but not compressed. By pressurizing a fluid in a closed system, the fluid can transmit both motion and force. This is accomplished by using pistons in cylinders of different sizes to either increase force or increase movement. Whenever there is an increase in force, there will be a decrease in movement. Conversely, when there is an increase in movement, there will be a decrease in force. In this manner, a hydraulic system is like the lever. Locate two items around your lab that use hydraulics to operate.

1. Item one: _____

    How does this item use hydraulics? _____

2. Item two: _____

    How does this item use hydraulics? _____

A simple hydraulic system contains two equal-sized containers of a liquid with two equal-sized pistons. The two containers are connected with a hose or tubing. An example of this is shown in Figure 4-21. If one piston is pushed downward with 100 lb (45 kg) of force and moves down 10 inches, the piston in the second container will move upward 10 inches with the same 100 lb (45 kg) of force. Since the pistons are the same size, any force and movement imparted on one piston will cause the same reaction to the second piston.

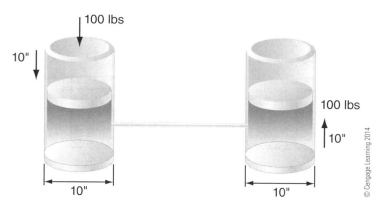

Figure 4-21

Where the use of hydraulics really provides advantage is when the sizes of the pistons are different; the resulting force and movement can be increased or decreased as needed. The pressure generated by the piston is a factor of the piston size. Input pressure is found by dividing force by piston size, or $P = F/A$. The smaller the input piston surface area, the larger the force will be from that piston. The larger the input piston surface area, the less the force will be from that piston, as shown in Figure 4-22. Conversely, the output piston force is proportional to the pressure against the surface area of the piston. An output piston that is larger than the input piston will move with greater force than the input but with less distance moved. To examine this principle we will look at what is known as Pascal's principle or law.

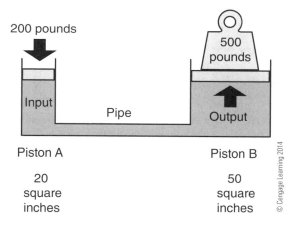

Figure 4-22

In Figure 4-23, the smaller piston on the left is our input piston and has a surface area of 1 square inch (6.45 cm²), and the larger piston is our output piston and has an area of 10 square inches (65.5 cm²). A force of 100 lb (45 kg) is exerted on the input piston, moving it downward 1 inch (2.54 cm). Using $P = F/A$, we can calculate $P = 100/1$, or 100 psi. Our output piston, which has a surface area of 10 square inches, will receive 100 pounds of pressure per square inch. This will result in an output force by piston 2 of 1,000 lb. However, since our output force has increased, our output movement will decrease. The larger piston will move one-tenth of the distance of the input piston; since the force was multiplied by 10, the distance will be divided by 10 as well. The 1 inch of movement of piston 1 turns into 1/10 of an inch of movement at piston 2. To calculate the forces and movement in a hydraulic circuit, use the following formulas:

$P_1 = F_1/A_1$ for piston 1 pressure

$F_2 = A_2/A_1 \cdot F_1$ for piston 2 force

$P_2 = F_2/A_2$ for piston 2 pressure

To practice using these principles, complete the following activity.

INSTRUCTOR VERIFICATION:

84   Chapter 4 Basic Technician Skills

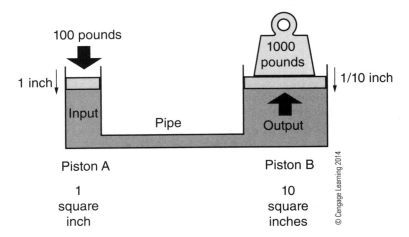

Figure 4-23

3. Calculate the forces and movements of a two-piston hydraulic circuit with an input piston of 2 square inches and an output piston of 10 square inches and an input force of 500 pounds (227 kg).

   Input distance _____ Input force _____

   Output distance _____ Output force _____

4. Calculate the forces and movements of a two-piston hydraulic circuit with an input piston of 2 square inches and an output piston of 1 square inch and an input force of 800 pounds (364 kg).

   Input distance _____ Input force _____

   Output distance _____ Output force _____

Obtain a selection of syringes from your instructor. Fill each syringe approximately halfway with water. Connect a hose between two different syringes. Push the plunger of one syringe, and note the reaction of the other. Perform this exercise several times using different sizes of syringes as the input and output pistons.

5. Syringe 1 diameter _____ Distance moved _____

   Syringe 2 diameter _____ Distance moved _____

6. Syringe 1 diameter _____ Distance moved _____

   Syringe 2 diameter _____ Distance moved _____

7. Syringe 1 diameter _____ Distance moved _____

   Syringe 2 diameter _____ Distance moved _____

8. Syringe 1 diameter _____ Distance moved _____

   Syringe 2 diameter _____ Distance moved _____

F. Electricity

Mostly everyone is familiar with two very common electrical occurrences, lightning, and static electricity. Lightning is an electrical discharge due to static electricity, just like when you get shocked touching a doorknob on a dry winter day. The difference is the energy in a lightning bolt is many, many times greater.

1. Describe what you think are the conditions necessary for getting a static electricity shock. _____
   _____

2. Why do you think a lightning discharge is so much more powerful than a static electricity shock? _____
   _____

The term *static electricity* refers to a buildup of electrical charges where there is typically poor electrical conductivity. When there is an electrical charge, there is the potential for electron flow. When an electron leaves an atom, it leaves behind an opening. The atom from which the electron left will now have a positive charge due to the loss of the negatively charged electron. The atom the electron moved to now has a negative charge. Whenever an imbalance occurs, there is the potential for electrical current flow. When an electrical force is present, and electrons move from atom to atom, electrical current flow exists. If a material does not easily accept the movement of electrons, it is an insulator or a poor conductor. Metals such as copper, silver, gold, and aluminum can readily give up and transfer electrons, and are considered good electrical conductors. Figure 4-24 shows how electrons move between atoms.

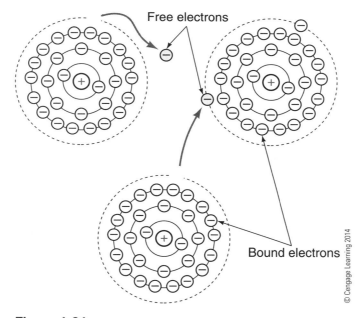

Figure 4-24

If during this static electricity transfer, dissimilar charges are present, the charges will attract. If the charges are similar, the charges will repel. Just as two magnets will repel when north poles are placed close together and attract if the north and south poles are near each other. If you have ever pulled a wool sweater on over your head and had your hair stand on end, you have experienced this phenomenon. When you pulled the sweater over your head, electrons transferred from the wool to your hair. Similarly, rubbing a balloon against a sweater can cause enough of a static charge to allow the balloon to "stick" to a wall.

3. Inflate two balloons and attach a length of string to each knotted end. Rub one balloon against a piece of clothing. Then hold the string so that the balloons hang down near each other. Describe what takes place.
   _____

4. Rub the balloons against your clothing and attempt to stick them to various surfaces. What surfaces stuck the best? _____ Which were the worst? _____

5. Describe why you think some surfaces hold the balloons better or worse than others. _____

6. What environmental factors affect static electricity? _____

To discuss electricity it is necessary to include magnetism because the two are interrelated and inseparable. Electromagnetism refers to the interaction of electricity and magnetism. When a current flows through a conductor, a magnetic field develops around the conductor. Likewise, when a conductor is moved through a magnetic field, current is induced into the conductor. This means that electricity can be generated by moving a conductor through a magnetic field. It is this principle that is the basis of electrical power generation, whether under the hood of a car with a generator or at a power generation facility, like a hydroelectric dam or a nuclear power plant. Electromagnetism also defines how relays, electric motors, and ignition coils operate.

Obtain a magnet and some iron shavings from your instructor. Place a piece of paper on top of the magnet, then sprinkle some of the shavings on the paper over the magnet.

7. What shapes do the shavings show on the paper? _____

8. Describe why the shavings appear in the patterns you see. _____
_____

Ask your instructor for a magnetic compass. To use the compass as an electrical current detector, place it along the positive or negative battery cable of a vehicle. Place the compass so the needle is parallel with the cable.

9. Turn on the headlights and describe the reaction of the compass. _____

10. Why did the compass react the way it did? _____

11. How would the compass react to a smaller electrical load than the headlights? _____

As you have demonstrated, electrical current flow generates magnetic fields. In Chapter 17 you will see how magnetic fields are used in motors and to generate electricity to charge the battery and power the vehicle's electrical system.

G. Pressure and Vacuum

Air pressure, or atmospheric pressure, is a measurement of the weight of the air from the edge of the atmosphere down to ground level. We do not often think about the air that surrounds us, but the weight of that air has a huge impact on our lives. Air pressure at sea level is about 14.7 psi. This means that if we could take a 1-square-inch column of air from the ground up to the edge of space, it would have a weight of approximately 14.7 pounds, as shown in Figure 4-25. As you move higher and higher above sea level, air pressure decreases.

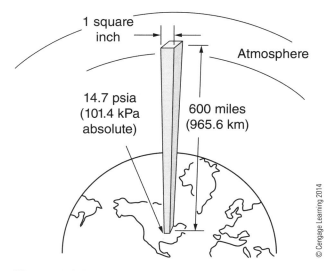

Figure 4-25

1. Explain why air pressure decreases with altitude. _____

   Since air pressure is a product of our atmosphere, it is subject to changes in the atmosphere. In addition, since air pressure is affected by the weather, it can be easily measured and recorded in your lab on a daily basis by making this simple barometer. Take a glass container and stretch a balloon or similar material over the opening of the jar. If necessary, use a rubber band to seal the balloon around the jar: it must be airtight. Next, attach a straw horizontally to the balloon with a piece of tape. The height of the end of the straw can be marked on a piece of paper to record the changes in air pressure. As the local air pressure changes, the straw will rise or drop with the pressure changes. Record the height over several days or weeks to track the changes in air pressure.

2. Describe how this simple barometer operates. _____

3. What causes the up or down movement of the straw? _____

4. If the air in the jar were heated or cooled, how would the straw react? _____

5. How does temperature affect the pressure inside the jar? _____

   We saw in the last activity that the air pressure changes, but what can cause the pressure to change? Temperature is one factor that can cause pressure change. When air is heated, it rises, and when air is cooled, it falls. The reason for this is that as air is heated, it weighs less per unit of volume than cooled air. Since the hotter air is less dense, the cooler, denser air lifts it. To demonstrate how temperature affects pressure, fill a plastic milk jug approximately one-third of the way full with very hot water. Use caution when working with hot water to avoid being accidentally burned. Screw the cap on securely, and wait about an hour.

6. What happened to the jug? _____

7. Why did the jug react in the way it did? _____

8. How did the outside air pressure affect the jug? _____

Chapter 4 Basic Technician Skills

To see how increasing the air temperature will result in an increase in air pressure, place a balloon over the opening of a plastic water bottle that is filled about halfway. Carefully heat the water bottle with a heat gun, hair dryer, or other safe and approved method.

9. Describe what happened to the balloon. _____

10. Why did the balloon react the way it did? _____

11. What does the balloon do as the water and bottle slowly cool? _____

12. Why did this happen? _____

The pressure of air (or a gas) in a container can be affected by several things, such as the temperature in the container and the volume of the container. In an automotive engine, air and fuel is drawn into the cylinder, which once sealed, becomes smaller. This decrease in volume causes the pressure and temperature of the gas to increase as the molecules are forced more tightly together.

When the volume of the cylinder is increased, such as when the piston moves from TDC to BDC on the intake stroke, the pressure in the cylinder decreases.

13. Why does the pressure decrease? _____

14. What causes the air to enter the cylinder? _____

Place the small open end of a syringe against one of your fingers and draw the plunger out until it stops moving.

15. Describe what happens. _____

16. What was the effect within the syringe by pulling out on the plunger? _____

Attach a pressure/vacuum gauge to a syringe with the plunger about halfway pulled out. Pull the plunger back to the end of its travel and note the reading on the gauge.

17. Gauge reading _____

18. With the plunger at the end of its travel, push the plunger in and note the reading on the gauge. _____
_____

Obtain two syringes of equal size and pull each plunger out about halfway. Then connect the two together with a piece of vacuum hose.

19. Push on one plunger and note what happens to the other plunger. _____

20. Was the result what you expected? Why or why not? _____

Chapter 4 Basic Technician Skills   89

21. What caused the movement of the second plunger? _____

22. Now, pull on the plunger and note what happens. _____

23. Why did the second plunger move? _____

24. How did the change in pressure in the first syringe affect the second syringe? _____
Obtain a vacuum pump and a cone-shaped dispensing lid from a differential fluid bottle from your instructor. Next, attach the cap to an empty motor oil quart bottle. Pull vacuum on the bottle via the cap and watch the oil bottle.

25. What happens to the oil bottle? _____

26. Why did the bottle react the way it did? _____

27. Release the vacuum and note the effect on the bottle. Explain the result. _____

Using a vacuum pump and a vacuum chamber connected to a bottle of oil, apply vacuum and watch the oil.

28. Why did the oil move from the oil bottle? _____

You have now demonstrated a basic physical principle; that pressure moves from a higher pressure to a lower pressure.
Next, place a balloon into an empty water bottle and leave the balloon's opening above the opening of the bottle. Now try to inflate a balloon that is inside the bottle.

29. Explain what happened. _____

30. Why did the balloon not inflate normally? _____

31. What could be done to allow the balloon to properly inflate? _____

The balloon was trying to inflate within the bottle and was exerting pressure against the air already inside the bottle. Since the air in the bottle had nowhere to go, the pressure equalized against the pressure you tried to place in the balloon, causing a pressure balance. The harder you tried to blow up the balloon, the more the pressure in the bottle worked against you. Now place a small hole in the bottom of the bottle and repeat the experiment.

32. Describe what happened. _____

33. Why was the balloon able to inflate more than before? _____

34. If you plug the hole in the bottle and release the air from the balloon, what will happen to the bottle? _____
_____

35. Why? _____

You have seen how pressure can move from high to low and how to increase pressure. Next, we will examine low pressure, also called vacuum. Vacuum is pressure that is less than atmospheric pressure. Pressure can be measured with a pressure gauge, such as a tire pressure gauge, but a tire gauge is calibrated to read atmospheric pressure as zero. This is referred to as psi gauge, or psig. A flat tire will read zero psi on a pressure gauge, but it actually has 14.7 psi (or equivalent local) pressure. To read pressures lower than atmospheric, you will need either a gauge that reads psi absolute (psia) or a vacuum gauge. Using a vacuum gauge and a pressure chart, you can easily see what a vacuum gauge reading reflects in actual air pressure. Connect a vacuum gauge to a vacuum hose and start the engine.

36. Record the gauge reading. _____

37. What does this equate to in actual air pressure? _____

38. Why is the engine producing vacuum? _____

39. What does the vacuum gauge read if the engine speed is increased? _____

H. Sound

Sound is a result of vibrations, though some vibrations can occur at frequencies either below or above the human range of hearing. As an object vibrates, it moves the air around the object, causing the air itself to vibrate. These vibrations are detected by the bones in our inner ears, producing the sounds that we hear. Frequency is the number of times a vibration occurs and is usually rated in Hertz (Hz), which is the number of cycles per second, shown in Figure 4-26. A sound with a vibration of 200 Hz will have 200 oscillations per second. A sound with a frequency of 2 KHz (2,000 Hz) will oscillate 2,000 times per second. The lower the Hertz rate, the lower the frequency and the tone of the sound. The higher the frequency, the higher pitch of the sound will be. Humans can usually hear sounds between 20 Hz and 20,000 Hz. As an example, the low string on a bass guitar vibrates at about 41 Hz, and the low string on a six-string guitar vibrates twice as fast at 82 Hz. The highest note on a six-string guitar vibrates around 1,300 Hz.

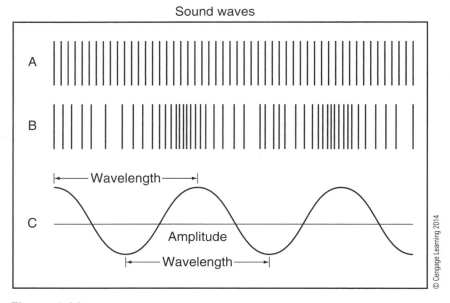

Figure 4-26

Being able to identify various sounds is an important diagnostic skill. By listening to a noise and its relation to engine speed or wheel speed, you can more easily locate the source of the problem.

1. A customer complains there is a high-pitched rattling noise from the vehicle that gets louder and increases in frequency when engine speed is increased. The noise occurs in Park and while driving. Which of these is the most likely cause and why?

    a. Loose wheel bearings

    b. Loose drive belt

    c. Loose exhaust shield

    d. Loose motor mount

    Why? _____

2. A customer states that a low-sounding rumble or roar occurs as the car is moving. The noise increases in pitch (frequency) as speed increases. Which of these is the most likely cause and why?

    a. Worn wheel bearing(s)

    b. Worn drive belt

    c. Broken exhaust shield

    d. Loose motor mount

    Why? _____

3. A customer brings her car in for a high-pitched squealing sound when the car is moving. The noise changes slightly when the brakes are applied and stops when the car stops moving. Which of these is the most likely cause and why?

    a. Worn wheel bearing(s)

    b. Brake pad wear indicator

    c. Broken exhaust pipe

    d. Loose drive belt

    Why? _____

X. Basic Chemistry

1. List two examples of chemical reactions that take place in the automobile.

    _____
    _____

2. Based on one of the examples you listed above, describe in detail what happens during the process, including what causes the reaction to take place, what factors affect the reaction, and what changes take place from the start to the finish of the reaction.

    _____
    _____
    _____
    _____

Lab Worksheet 4-1

Name _____ Date _____ Instructor _____

Communication Skills

Contact a local parts supplier and obtain the following information necessary to complete an estimate.

Year _____ Make _____ Model _____

VIN _____

Name of parts store _____

Name of parts person taking the call _____

Use the following script to find the cost and availability of the parts:
   Hello, this is (your name here) calling from the automotive program at (school name here). I need our cost on the following parts and their availability.

   a. Part _____   Cost _____
   b. Part _____   Cost _____
   c. Part _____   Cost _____
   d. Part _____   Cost _____
   e. Availability of the parts _____

   Record the cost and availability in the spaces. Once you have the parts costs and their availability, thank the person on the phone and tell him or her that you will call back once you find out if you will need the parts.

INSTRUCTOR VERIFICATION:

Lab Worksheet 4-2

Name _____ Date _____ Instructor _____

Using Service Information

Record the identifying information for a vehicle, then locate the service information that follows:

Year _____ Make _____ Model _____

VIN _____ Engine _____

1. Wheel lug torque spec _____

2. Firing order _____

3. Drive belt routing (draw picture)

4. Engine oil capacity _____

5. Recommended coolant type _____

6. Labor time to replace front brake pads _____

7. Labor time to replace the fuel pump _____

8. Locate a technical service bulletin (TSB) of your choice; record the TSB number, TSB date, and a brief description of the reason for the TSB.

INSTRUCTOR VERIFICATION:

CHAPTER 5

Wheels, Tires, and Wheel Bearings

Review Questions

1. Tires ___Support___ the weight of the vehicle, absorb much of the road ___imperfections___, and are vital to how the vehicle handles and how well it ___steers/maneuvers___.

2. Tires that rely on air pressure are called:
   a. Hydraulic.
   (b) Pneumatic.
   c. Aquatic.
   d. Electronic.

3. True or (False) With the weight of the vehicle on the tire, the tire will remain perfectly round.

4. A tire contact patch of 36 square inches at 28psi can support how much weight?
   a. 960 pounds
   b. 1255 pounds
   (c) 1008 pounds
   d. 888 pounds

5. Explain why temporary spare tires require higher air pressure than the standard size tires?
   ___As the contact patch is smaller, higher pressure is required to support the vehicle's weight___

6. When the weight of the vehicle is on the tires, the tires tend to ___flex/bulge___ slightly.

7. When the temperature of the tire increases, such as during driving, the air pressure inside the tire ___increases___.

8. The type of tire balance shown in Figure 5-1 is:

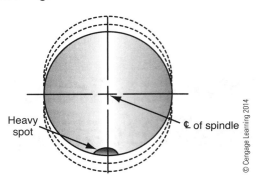

Figure 5-1

   a. Dynamic.
   b. Lateral.
   c. Radial.
   (d.) Static.

9. Dynamic wheel imbalance causes the steering wheel to shake or __shimmy__.

10. Technician A says radial tires are the most common tire type used today. Technician B says radial tires use layers of materials such as steel and polyester in their construction. Who is correct?

   a. Technician A
   b. Technician B
   (c.) Both A and B
   d. Neither A nor B

11. In general, the shorter a tire's sidewall is, the __stiffer__ the tire will be.

12. Identify the tire types shown in Figure 5-2 (a through c).

   a. __Assymetrical__

b. ____Directional____

c. ____Mud + Snow____

Figure 5-2

13. List four different types of tires.
    ____All season, Winter, Summer, All weather____

14. The majority of tires used on passenger vehicles are ____All____ - season tires.

15. Sports car tires may not perform well in ____Colder____ temperature operation.

16. An _asymmetrical_ tire is one that has different tread patterns on the inside and outside sections of the tread.

17. A _directional_ tire is one that has a tread pattern that is made to rotate primarily in one direction.

18. Most temporary spare tires require an inflation pressure of _60_ psi.

19. Self-supporting run-flat tires require the use of a tire _pressure_ monitoring system.

20. Which of the following is not part of the tire size marking on a tire?
    a. Rim diameter
    b. Rim width
    (c.) Tread section width
    d. Aspect ratio

21. The _temperature_ and _traction_ ratings use letters, from C to A or AA to indicate overall tire quality.

22. Most passenger car tires have maximum inflation pressures between _35_ psi and _45_ psi.

23. The last four digits of the _DOT_ number provide the production date of the tire.

24. List three types of tire problems that can result from tire plies that are not molded correctly.
    Separation, Balance, Radial force Variation

25. Tire pull can be caused by _____, meaning the tire is slightly cone shaped.

26. Steel wheels are often covered with _____ to enhance the look of the vehicle.

27. A hub-centric wheel means that the wheel is centered to the vehicle by the:
    a. Lug nuts.
    (b.) Hub.
    c. Hub cap.
    d. All of the above.

28. Technician A says changing wheel offset can affect wheel bearing life. Technician B says changing wheel offset can cause the tire to interfere with the brake system. Who is correct?
    a. Technician A
    b. Technician B
    (c.) Both A and B
    d. Neither A nor B

29. Technician A says overtightening lug nuts can damage the hub. Technician B says overtightening the lug nuts can damage the brake rotors. Who is correct?
    a. Technician A
    b. Technician B
    c. Both A and B
    d. Neither A nor B

30. The two types of tire pressure monitoring systems (TPMS) are:
    a. Static and dynamic.
    b. Direct and indirect.
    c. Lateral and radial.
    d. Internal and external.

31. Explain the differences between the two types of tire pressure monitoring systems.

    Direct - pressure sensor mounted inside tire
    Indirect - pressure is determined by tire rotation

32. Describe why it is important to use caution when inspecting a tire.

    There could be foreign objects imbedded in the treads, cords may have worn through, or tire pressure may be high enough for explosion.

33. Explain how to use a tire machine to dismount and remount a tire on a wheel.

    _____
    _____
    _____
    _____

34. Tire pressure should be set when the tire is __cold__. If tire pressure is checked when the tire is hot, the pressure may read __higher__.

35. Draw the two common patterns for performing a tire rotation.

    Diagonal   Front to Back

36. Explain what is indicated by finding rubber dust inside of the tire.

    Dirt inside tire when mounting or low tire pressure.

37. If a wheel and tire are stuck on the hub, describe what steps should be taken to break the wheel loose from the hub.

38. Which of following is not a common reason for tire air loss?
    a. Debris punctured the tread.
    b. Rust or corrosion around the rim and bead.
    c. Cracks in the rim.
    d. Leaking valve stem or core.

39. A customer complains that the vehicle's steering wheel shakes while driving on the freeway but is fine at lower speeds. What is the most likely cause?
    a. Loose wheel bearings
    b. Tire imbalance
    c. Tire conicity
    d. Low tire pressure

40. A wheel/tire that is __statically__ out of balance will try to bounce or hop as it is driven.

41. Technician A says excessive wheel/tire runout can cause a vibration. Technician B says a bent hub can cause a vibration. Who is correct?
    a. Technician A
    b. Technician B
    c. Both A and B
    d. Neither A nor B

42. A vehicle has an illuminated TPMS light on the dash. Technician A says one or more tires may have low air pressure. Technician B says if the TPMS light stays on with the tires inflated correctly it means that there is a fault in the system. Who is correct?
    a. Technician A
    b. Technician B
    c. Both A and B
    d. Neither A nor B

43. Explain how bearings, particularly ball and roller bearings, reduce friction.

    By increasing the surface area

44. Technician A says all FWD vehicles use a sealed wheel bearing and hub assembly that is replaced as a unit. Technician B says some FWD vehicles use tapered roller bearings for the front wheel bearings. Who is correct?

    a. Technician A
    b. Technician B ✓
    c. Both A and B
    d. Neither A nor B

Activities

1. For a vehicle specified by the instructor, use a ruler to determine the approximate contact patch area for a tire sitting on the floor and at its proper inflation pressure.
   a. Contact patch width _____ Length _____ Total area _____
   b. Multiply the tire pressure by the contact patch area to determine how much weight can be carried.
      _____ lb (kg)
   c. Record the maximum load rating specified on the tire sidewall. _____
   d. Find the difference between the specified maximum load and the weight you determined in question b.
      _____
   e. If there is a difference in the two numbers, why? _____
   f. How does inflation pressure affect the load carrying capacity of the tire? _____
   g. How does tire size (width) affect tire load carrying capacity? _____
   h. Determine if the vehicle has a temporary spare tire; if so, what is the recommended pressure for this tire?
      _____
   i. Why is the tire pressure for a mini-spare higher than the pressure for the regular tires?
      _____

2. Tire and wheel construction
   a. Label the parts of the tire shown in Figure 5-3 (a through c).

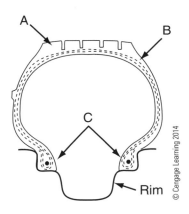

Figure 5-3

b. Label the tire dimensions in Figure 5-4.

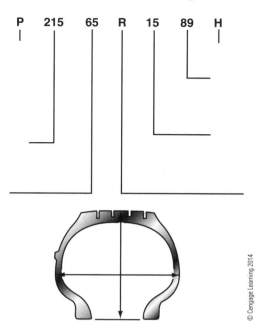

Figure 5-4

c. Label the parts of the wheel shown in Figure 5-5 (a through h).

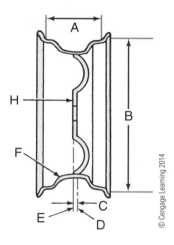

Figure 5-5

d. Match the tire sidewall information with its description. Passenger car tire

Section width	93
Aspect ratio	P
Rim diameter	18
DOT number	235
Speed rating	40
Load index	MXTB A9A 1511

106 Chapter 5 Wheels, Tires, and Wheel Bearings

3. Wheel bearing

    a. Label the types of wheel bearings shown in Figure 5-6 (a through c).

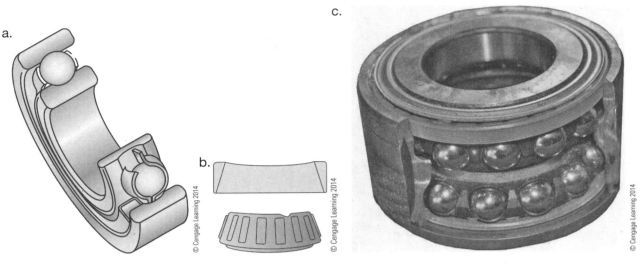

Figure 5-6

b. Label the parts of the wheel bearing shown in Figure 5-7 (a through c).

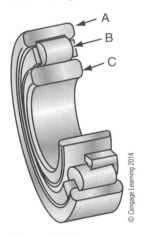

Figure 5-7

c. Label the forces that act upon wheel bearings during operation in Figure 5-8 (a through c).

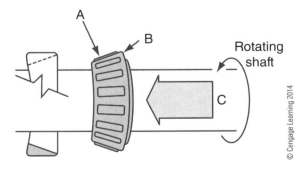

Figure 5-8

INSTRUCTOR VERIFICATION:

Lab Worksheet 5-1

Name: Shane Silbernagel Date: 2016/02/29 Instructor: ____

Tire Identification

Locate the following tire information:

Vehicle year: 2012
Tire manufacturer: BF Goodrich
Tire size: P245/55 R18
Traction rating: A
Load rating: 850 kg / 1874 Lbs.

Vehicle make and model: Chev Camero
Tire model: Radial T/A
Treadwear rating: 620
Temperature rating: B
Speed rating: T

Vehicle year: 2007
Tire manufacturer: Dunlop
Tire size: P265/70 R17
Traction rating: B
Load rating: 1150 kg / 2535 Lbs.

Vehicle make and model: Toyota FJ Cruiser
Tire model: AT20
Treadwear rating: 300
Temperature rating: B
Speed rating: S

Vehicle year: 2009
Tire manufacturer: Hankook
Tire size: P195/60 R15
Traction rating: B
Load rating: 540 kg / 1190 Lbs.

Vehicle make and model: Ford Focus
Tire model: Optimo
Treadwear rating: 620
Temperature rating: B
Speed rating: T

INSTRUCTOR VERIFICATION:

Lab Worksheet 5-2

Name: Shane Silbernagel Date: 2016/02/29 Instructor: _____

Vehicle year: 2009 Vehicle make and model: ~~Subaru Outback~~ Ford Focus

Checking Tire Pressure

Tire pressure spec: 32 psi   Spare pressure spec: 60 psi
RF tire pressure: 23 psi   LF tire pressure: 23.5 psi
RR tire pressure: 24 psi   LR tire pressure: 25 psi

Vehicle year: 2006   Vehicle make and model: Toyota FJ
Tire pressure spec: 32   Spare pressure spec: 32 psi
RF tire pressure: 28.5   LF tire pressure: 28
RR tire pressure: 28.5   LR tire pressure: 29.5

Vehicle year: 2012   Vehicle make and model: Chev. Camaro
Tire pressure spec: 35 psi   Spare pressure spec: 60 psi
RF tire pressure: 25   LF tire pressure: 37.5
RR tire pressure: 26   LR tire pressure: 40.5

INSTRUCTOR VERIFICATION:

Lab Worksheet 5-3

Name _____ Date _____ Instructor _____

Checking Tire Wear Patterns

Using the tire wear patterns shown in Figure 5-9, determine any wear concerns for the tires specified by your instructor.

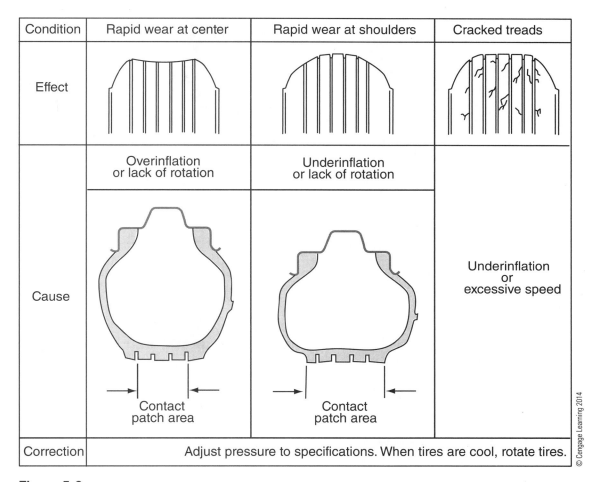

Figure 5-9

Vehicle year _____ Vehicle make and model _____
RF tire wear _____ LF tire wear _____
RR tire wear _____ LR tire wear _____

Vehicle year _____ Vehicle make and model _____

RF tire wear _____ LF tire wear _____

RR tire wear _____ LR tire wear _____

Vehicle year _____ Vehicle make and model _____

RF tire wear _____ LF tire wear _____

RR tire wear _____ LR tire wear _____

INSTRUCTOR VERIFICATION:

Lab Worksheet 5-4

Name _____ Date _____ Instructor _____

Year _____ Make _____ Model _____

Tire Rotation

1. Find the vehicle lift points and support correctly. Vehicle supported where and how? Include diagram of locations. _____

2. Locate manufacturer-recommended tire rotation pattern, and include a diagram of pattern, or draw and describe. _____

3. Check all wheels and tires for the following:
   Wheel damage, cuts or gouges, abrasions, center cap condition _____
   Tire damage, tread and sidewall _____
   Tire wear _____
   Valve stem damage _____
   Incorrect inflation _____
   Lug nut condition _____

4. Remove wheels and relocate to specified location. Note any issues of excessive rust or corrosion between the wheel and the hub. _____

5. Recommended tire pressure spec F _____ R _____ Set tire pressure to manufacturer specifications.

6. Reinstall wheels and torque lug nuts to spec. Torque spec _____ and pattern _____

7. Final visual inspection and instructor sign-off.

   _____ Date _____

INSTRUCTOR VERIFICATION:

Lab Worksheet 5-5

Name _____ Date _____ Instructor _____

Tire size _____ Tire pressure _____

Balance Wheel and Tire Assembly

1. Inspect the wheel and tire for any visible damage and note your findings. _____
   _____

2. What types of damage to the wheel and/or tire will affect wheel/tire balance? _____
   _____

3. Remove the center cap if necessary and mount the wheel/tire to the balancer. Spin the tire by hand and note any excessive lateral (side-to-side) or radial (up/down) movement. _____ Excessive lateral or radial movement may indicate a problem with the wheel/tire or may indicate improper mounting on the balancer. Double-check that the wheel/tire is correctly mounted to the balancer before continuing.

4. Input the wheel/tire information into the balancer as necessary.
   Wheel diameter _____
   Wheel width _____
   Wheel offset _____

5. Start the balancer and note the required weight correction amounts.
   Inside weight _____   Outside weight _____
   If the wheel/tire is out of balance, remove any wheel weights and recheck.
   Inside weight _____   Outside weight _____

6. Install the correct weight and recheck tire balance. OK _____ Not OK _____
   Instructor check _____

7. If the wheel/tire is still out of balance, what actions are necessary to bring it into balance? _____
   _____

INSTRUCTOR VERIFICATION:

Lab Worksheet 5-6

Name _____ Date _____ Instructor _____

Year _____ Make _____ Model _____

Reinstall Wheel and Torque Lug Nuts

1. Locate and record the wheel lug nut torque specs. _____

2. Draw the recommended pattern in which the lug nuts are tightened.

3. Inspect the wheel studs and hub before setting the wheel in place. Note any signs of damage to the studs or corrosion on the hub. _____

4. Inspect the wheel lug bore and centerbore for damage; note your findings. _____
   _____

5. Inspect the lug nuts for internal damage to the thread (external threads if applicable for certain vehicles) and for external damage to the nut. Note your findings _____
   _____

6. If all studs and lug nuts are undamaged and usable, place the wheel onto the hub and studs. Start each lug nut by hand or using a socket by hand only. Seat each lug nut so that the wheel centers and seats properly on the hub.
   Instructor check _____

7. Set the torque wrench to the proper setting for the lugs.
   Instructor check _____

8. Tighten each lug in sequence and repeat to ensure all lugs are fully seated and tight.
   Instructor check _____

9. Reinstall hub caps or center caps as necessary.

INSTRUCTOR VERIFICATION: _____

Lab Worksheet 5-7

Name _____ Date _____ Instructor _____

Year _____ Make _____ Model _____

Tire size _____ Wheel type _____

Inspect Wheel and Tire for Air Loss

1. Check and record tire pressure.

   Recommended tire pressure

   If necessary, inflate tire to proper pressure before continuing.

2. Carefully inspect the tire tread section, sidewall, and wheel for signs of damage. Note your findings.
   _____
   _____

3. If your visual inspection does not reveal an obvious cause for air loss, use either a tire dip tank or a spray bottle with a soapy water solution to locate the leak. Note your findings.
   _____

4. If the tire has a puncture in the tread and it is within the repairable area, what type of repair should be made?
   _____

5. If the tire is leaking around the bead area, what actions are necessary to attempt to repair the wheel/tire?
   _____

6. If the leak is from or around the valve stem, what steps are required to repair the leak?
   _____

7. If the tire is leaking from sidewall damage or dry rot, what is required?
   _____

INSTRUCTOR VERIFICATION:

Lab Worksheet 5-8

Name _____ Date _____ Instructor _____

Year _____ Make _____ Model _____

TPMS tool _____ Number of TPMS sensors _____

Tire Pressure Monitoring System Service

1. Perform TPM system checks.
    a. TPMS light/indicator operates KOEO.   OK _____   Not OK _____
    b. Start engine and note TPMS light/indicator. _____
    c. What is indicated by this? _____
    d. Is actual tire pressure displayed?   Yes _____   No _____
    e. Record displayed readings.   RF _____ LF _____ RR _____ LR _____

2. Follow the tire pressure sensor activation sequence. If frequency data is not provided, record the temperature shown by the sensor.

3. LF sensor frequency and pressure _____   RF _____
    LR sensor frequency and pressure _____   RR _____
    Spare tire frequency and pressure _____

4. Compare the TPMS readings to a tire pressure gauge.
    Gauge readings: LF _____ RF _____ LR _____ RR _____
    Are the readings very close to each other?   Yes _____   No _____
    What could cause the readings to be different? _____

5. Summarize the manufacturer's TPMS reset procedure. _____
    _____
    _____

6. Does the TPMS require a relearn procedure for a tire rotation?   Yes _____   No _____
    If yes, describe the relearn procedure. _____

INSTRUCTOR VERIFICATION:

Lab Worksheet 5-9

Name _____ Date _____ Instructor _____

Year _____ Make _____ Model _____

Inspect Wheel Bearing

1. Describe the common symptoms of a faulty wheel bearing. _____
   _____
   _____

2. Does the bearing make noise when driving the vehicle?   Yes _____   No _____

3. If the bearing is making noise, does the noise change when turning a corner?
   Yes _____   No _____
   If yes, describe how the noise changes and what this indicates? _____
   _____

4. Raise and support the vehicle and spin the tire by hand. Note any roughness or looseness in the bearing. Note your findings. _____
   _____

5. Grasp the tire at the three o'clock and nine o'clock positions and shake the tire. Is any looseness felt?
   Yes _____   No _____
   If yes, is there movement in the wheel bearing?   Yes _____   No _____

6. Based on your inspection, what is the recommended action? _____
   _____

INSTRUCTOR VERIFICATION:

CHAPTER 6: Suspension System Principles

Review Questions

1. All of the suspension components must work together to provide the __ride__ quality and __handling__ characteristics expected by the driver and passengers.

2. Summarize six functions of the suspension system.
   a. Absorb shock
   b. Characteristics of handling
   c. Carry load.
   d. Cornering abilities
   e. Safety in braking maneuvers
   f. Mounting platform for steering system

3. The __weight__ of the vehicle is carried through the springs to the wheels and tires.

4. Vehicles with a body-over-frame design often have a __ladder__ shaped frame.

5. __Cross members / subframe__ are frame components that attach to the frame rails and carry the engine and transmission.

6. Rubber __bushings / cushions__ are used to isolate the frame from the body.

7. Most modern vehicles use a __space / unibody__ type frame.

8. An __independent__ suspension allows each wheel to move separately for the best ride quality.

9. The type of front suspension used on many heavy-duty vehicles is the:
   a. Independent.
   (b.) I-beam.
   c. 4WD.
   d. None of the above.

Chapter 6 Suspension System Principles

10. A dependent rear axle that drives the wheels is called a __live__ axle.

11. Used on the rear of many FWD cars, a __trailing__ axle is a type of dependent rear suspension.

12. A semi-independent rear axle is able to __articulate/move__ slightly, providing improved ride and handling.

13. During braking, as much as __80%__ percent of the weight is transferred to the front of the vehicle.

14. If the rear tires of a vehicle reach their grip and cornering limits before the front tires, it is called __oversteer__.

15. When the front of the vehicle cannot make a turn through the desired turn radius because the front tires have lost traction, it is called __understeer__.

16. When a coil spring __compresses__, it stores energy.

17. Coil springs often use rubber __bushings/cushions__ between the spring and the frame to reduce noise.

18. Describe the difference between how standard and variable-rate coil springs operate.
    Standard moves quite uniformly throughout its range to carry load. Variable are meant to provide more comfort through the ride while still

19. Leaf springs are attached to the frame with a __shackle__ or bracket.

20. Adding additional leaf springs allows the vehicle to carry __more__ weight.

21. Technician A says air springs are used only on large commercial trucks because they make the vehicle ride very rough. Technician B says air springs are adjustable, meaning ride height or load-carrying capacity can vary depending on the air pressure in the springs. Who is correct?
    a. Technician A
    b. Technician B
    c. Both A and B
    d. Neither A nor B

22. Torsion bars __twist__ when the control arm moves in response to road conditions.

23. Torsion bars are most commonly used on what type of vehicle?
    a. FWD
    b. AWD
    c. 4WD
    d. RWD

Chapter 6 Suspension System Principles

24. The upward movement of the tire is called _jounce_ and the downward movement of the tire is called _bounce_.

25. Technician A says that the weight carried by the springs is called sprung weight. Technician B says sprung weight includes the wheels and tires. Who is correct?
    - **a. Technician A**
    - b. Technician B
    - c. Both A and B
    - d. Neither A nor B

26. Shock absorbers are used to _dampen / control_ the spring oscillations.

27. As the shock piston moves up and down in the main chamber, oil is _transferred / directed_ into a second chamber.

28. Control arms are also called _long, short "A"_ -arms and _triangle / wishbone_ due to their similarity to those shapes.

29. Ball joints allow the steering knuckle to _move / turn_ while providing a tight connection to the control arms.

30. Describe the two types of ball joints. _Load carrying / stabilizing_

31. List four methods by which ball joints are attached to control arms. _riveted, bolted, threaded, or pressed._

32. Describe the purpose and operation of the stabilizer bar. _To minimize body roll and maintain contact patch of tire on road to provide more control/stability_

33. Explain the construction of a MacPherson strut. _Spring, shock and upper control arm_

34. A _modified_ strut locates the spring separate from the shock.

35. Explain how the strut differs between a multilink and MacPherson strut suspension.

36. List the components commonly used in a short/long arm suspension system.
   Control arms, Upper & lower ball joints, spring or torsion bar

37. A 4WD suspension may use a solid **live** axle or it may mount the differential to the chassis to allow for an **independent** front suspension.

38. List at least five components that are typically unsprung weight. *tires, wheels, ½ spring, rotors, calipers, steering knuckle, lower control arm.*

39. What components are eliminated by the MacPherson strut suspension?
   Upper control arm, ball joint

40. Technician A says the sprung weight of the vehicle is carried by the lower ball joint on a MacPherson strut suspension. Technician B says the sprung weight of the vehicle is carried by the lower ball joint on a modified strut suspension. Who is correct?
   a. Technician A
   (b.) Technician B
   c. Both A and B
   d. Neither A nor B

41. Many 4WD trucks and SUVs use **torsion bars** instead of coil springs.

42. Wind-up occurs when **torque** is applied to the rear axle.

43. Many rear suspension systems use a **torsion** bar or a **track** bar to limit rear axle movement.

44. Magneride uses a **magnetic** fluid that can change viscosity.

Activities

1. Identify the following suspension types.

   Figure 6-5 Figure 6-2 Figure 6-3 Figure 6-4 Figure 6-5
   SLA   MacPherson strut Modified strut Multilink Live axle
              Semi-independent axle   Torsion bar Independent 4WD

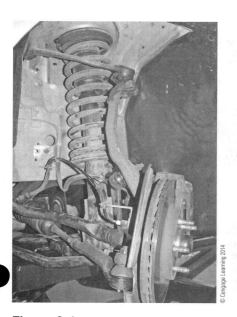

Figure 6-1

Figure 6-2

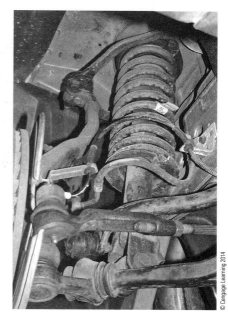

Figure 6-3

Figure 6-4

Figure 6-5

INSTRUCTOR VERIFICATION:

2. Identify the types of springs shown below.

Figure 6-6	Figure 6-7	Figure 6-8	Figure 6-9	Figure 6-10
Variable-rate coil	Standard-rate coil	Leaf	Torsion bar	Air

Figure 6-6

Figure 6-7

Figure 6-8

Figure 6-9

Figure 6-10

INSTRUCTOR VERIFICATION:

3. Circle the components that are the unsprung weight in Figure 6-11.

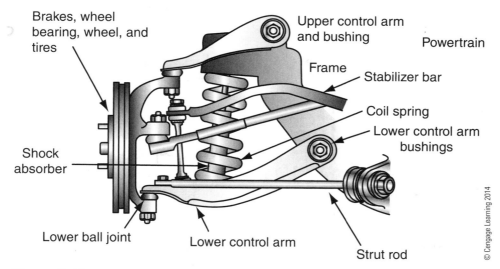

Figure 6-11

4. Label the load-carrying ball joint in the following figures.

Figure 6-12   Figure 6-13   Figure 6-14

SLA suspension   Modified strut suspension   Multilink suspension

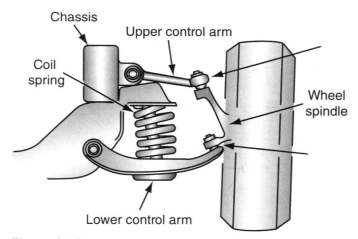

Figure 6-12

132 Chapter 6 Suspension System Principles

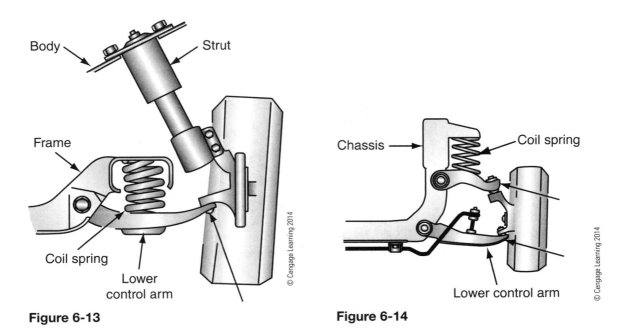

Figure 6-13

Figure 6-14

5. Label the components of the suspension shown below.

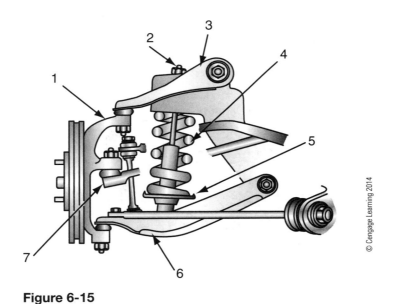

Figure 6-15

6. Match the component to the descriptions.

Component	Description
Spring	Connects the control arm to the stabilizer bar
Steering knuckle	Dampens spring oscillations
Shock absorber	Carries weight of vehicle
Upper control arm	Combines shock and spring
Lower control arm	Supports spring and load-carrying ball joint
Control arm bushing	Connects upper and lower control arms
Strut	Reduces body roll
Bearing plate	Typically carries following ball joint
Cam bolt	Allows strut to pivot
Strut rod	Used to adjust camber
Sway bar	Sometimes used to adjust caster
Sway bar links	Allows for control arm movement

INSTRUCTOR VERIFICATION:

Lab Worksheet 6-1

Name _____ Date _____ Instructor _____

Year _____ Make _____ Model _____

Identify Suspension Types and Components—Short/Long Arm.

1. Locate the control arms and describe the shape of each.
   Upper control arm _____
   Lower control arm _____

2. How many ball joints are used? _____

3. What is the location of the load-carrying ball joint? _____

4. How are the ball joints secured to the control arms?
   Upper ball joint _____
   Lower ball joint _____

5. What types of springs are used and where are they located? _____
   _____

6. Where are the shock absorbers mounted and how? _____
   _____

7. Describe the type of stabilizer bar links used. _____

8. How are changes to caster and camber made on this suspension? _____

9. Does the suspension use electronic ride control? _____ Yes _____ No
   If yes, describe the system and its basic operation. _____
   _____
   _____

10. Using service information, locate a TSB regarding the front suspension of this vehicle or of a vehicle with a similar suspension system. Record the TSB number, title, and a brief summary of the subject of the TSB.
    _____
    _____
    _____

INSTRUCTOR VERIFICATION:

Lab Worksheet 6-2

Name _____ Date _____ Instructor _____

Year _____ Make _____ Model _____

Identify Suspension Types and Components—MacPherson Strut

1. Locate the control arms and describe the shape of each.
   Lower control arm _____

2. How many ball joints are used? _____

3. How is the strut connected to the steering knuckle? _____

4. How are the ball joints secured to the control arms?
   Lower ball joint _____

5. What types of springs are used and where are they located? _____
   _____

6. Where are the shock absorbers mounted and how? _____
   _____

7. Describe the type of stabilizer bar links used. _____

8. How are changes to caster and camber made on this suspension? _____
   _____

9. Does the suspension use electronic ride control? _____ Yes   _____ No
   If yes, describe the system and its basic operation. _____
   _____
   _____

10. Using service information, locate a TSB regarding the front suspension of this vehicle or of a vehicle with a similar suspension system. Record the TSB number, title, and a brief summary of the subject of the TSB.
    _____
    _____
    _____

INSTRUCTOR VERIFICATION:

Lab Worksheet 6-3

Name _____ Date _____ Instructor _____

Year _____ Make _____ Model _____

Identify Suspension Types and Components—Multilink

1. Locate the control arms and describe the shape of each.
   Upper control arm _____
   Lower control arm _____

2. How many ball joints are used? _____

3. What is the location of the load-carrying ball joint? _____

4. How are the ball joints secured to the control arms?
   Upper ball joint _____
   Lower ball joint _____

5. What types of springs are used and where are they located? _____
   _____

6. Where are the shock absorbers mounted and how? _____
   _____

7. Describe the type of stabilizer bar links used. _____

8. How are changes to caster and camber made on this suspension? _____

9. Describe how the steering knuckle used in this suspension is different from steering knuckles used in MacPherson strut and SLA suspensions. _____
   _____

10. Does the suspension use electronic ride control? _____ Yes _____ No
    If yes, describe the system and its basic operation. _____
    _____
    _____

11. Using service information, locate a TSB regarding the front suspension of this vehicle or of a vehicle with a similar suspension system. Record the TSB number, title, and a brief summary of the subject of the TSB.

_____
_____
_____

INSTRUCTOR VERIFICATION:

Chapter 6 Suspension System Principles   141

Lab Worksheet 6-4

Name _____ Date _____ Instructor _____

Year _____ Make _____ Model _____

Identify Suspension Types and Components—I-Beam

1. Locate the I-beams and describe the shape of each.
   Left _____
   Right _____

2. How many ball joints are used? _____

3. What is the location of the load-carrying ball joint? _____

4. How are the ball joints secured to the steering knuckles?
   Upper ball joint _____
   Lower ball joint _____

5. What types of springs are used and where are they located? _____
   _____

6. Where are the shock absorbers mounted and how? _____
   _____

7. Describe the type of stabilizer bar links used. _____
   _____

8. How are changes to caster and camber made on this suspension? _____

9. Using service information, locate a TSB regarding the front suspension of this vehicle or of a vehicle with a similar suspension system. Record the TSB number, title, and a brief summary of the subject of the TSB.
   _____
   _____
   _____

INSTRUCTOR VERIFICATION:

© 2014 Cengage Learning. All Rights Reserved. May not be scanned, copied or duplicated, or posted to a publicly accessible web site, in whole or in part.

Lab Worksheet 6-5

Name _____ Date _____ Instructor _____

Year _____ Make _____ Model _____

Identify Suspension Types and Components—4WD

1. Describe the type of front suspension used on this vehicle. _____
   _____
   _____

2. Does this suspension use control arms? _____ Yes _____ No

3. How many ball joints are used? _____

4. What is the location of the load-carrying ball joint, if applicable? _____
   _____

5. How are the ball joints secured to the control arms?
   Upper ball joint _____
   Lower ball joint _____

6. What types of springs are used and where are they located? _____
   _____

7. Where are the shock absorbers mounted and how? _____
   _____

8. Describe the type of stabilizer bar links used. _____
   _____

9. How are changes to caster and camber made on this suspension? _____

10. Using service information, locate a TSB regarding the front suspension of this vehicle or of a vehicle with a similar suspension system. Record the TSB number, title, and a brief summary of the subject of the TSB.
    _____
    _____
    _____

INSTRUCTOR VERIFICATION:

Lab Worksheet 6-6

Name _____ Date _____ Instructor _____

Year _____ Make _____ Model _____

Identify Suspension Types and Components—Modified Strut

1. Locate the control arms and describe the shape of each.
   Lower control arm _____

2. How is the strut connected to the steering knuckle? _____
   _____

3. How are the ball joints secured to the control arms?
   Lower ball joint _____

4. Describe the type of stabilizer bar links used. _____

5. How are changes to caster and camber made on this suspension? _____

6. Does the suspension use electronic ride control? _____ Yes _____ No
   If yes, describe the system and its basic operation. _____
   _____
   _____

7. Using service information, locate a TSB regarding the front suspension of this vehicle or of a vehicle with a similar suspension system. Record the TSB number, title, and a brief summary of the subject of the TSB.
   _____
   _____
   _____

INSTRUCTOR VERIFICATION:

Lab Worksheet 6-7

Name _____ Date _____ Instructor _____

Year _____ Make _____ Model _____

Identify Suspension Types and Components—Rear Suspension

1. What type of rear suspension system is used on this vehicle? _____
   _____

Check all components that apply:
   Upper control arm _____
   Lower control arm _____
   Strut _____

2. How many ball joints are used? _____

3. What is the location of the load-carrying ball joint? _____

4. How are the ball joints secured to the control arms?
   Upper ball joint _____
   Lower ball joint _____

5. What types of springs are used and where are they located? _____
   _____

6. Where are the shock absorbers mounted and how? _____
   _____

7. Describe the type of stabilizer bar links used. _____

8. How are changes to caster and camber made on this suspension? _____

9. Does the suspension use electronic ride control? _____ Yes _____ No
   If yes, describe the system and its basic operation. _____
   _____
   _____

148   Chapter 6 Suspension System Principles

10. Using service information, locate a TSB regarding the rear suspension of this vehicle or of a vehicle with a similar suspension system. Record the TSB number, title, and a brief summary of the subject of the TSB.

   _____
   _____
   _____

INSTRUCTOR VERIFICATION:

CHAPTER 7

Suspension System Service

Review Questions

1. Describe five tools commonly used in suspension system service and repair.
   a. _____
   b. _____
   c. _____
   d. _____
   e. _____

2. Always use your _____ muscles when lifting.

3. Working on suspension components often means working with _____ components and fasteners.

4. Before beginning any work, make sure the vehicle is properly raised and _____.

5. Before using a strut compressor, make sure you are fully _____ in its use and understand how to safely operate the equipment.

6. _____ should never be applied to a shock absorber since it can cause the shock to rupture.

7. Inspection of the suspension system also includes looking at the _____ and the _____ system.

8. In every aspect of automotive service and repair, you first should _____ the customer's complaint.

9. Before any repairs can be made, a thorough _____ of the suspension system must be performed.

10. As the springs age and weaken, vehicle _____ _____ decreases.

11. Describe how vehicle ride and handling are affected by weak springs. _____
_____
_____

12. Explain how to check vehicle ride height. _____
_____

13. Technician A says weak shocks will affect ride height. Technician B says weak shocks will affect ride quality. Who is correct?
    a. Technician A
    b. Technician B
    c. Both A and B
    d. Neither A nor B

14. Technician A says installing smaller wheels and tires will affect ride height. Technician B says low tire pressure will affect ride height. Who is correct?
    a. Technician A
    b. Technician B
    c. Both A and B
    d. Neither A nor B

15. Which of the following is not a symptom of a worn-out shock?
    a. Noise when bouncing
    b. Oil leakage
    c. Excessive bouncing
    d. Incorrect ride height

16. True or False: A vehicle may show more than one type of tire wear. _____

17. A tire shows excessive wear in the center of the tread. Which is the most likely cause?
    a. Underinflation
    b. Overinflation
    c. Negative camber
    d. Positive camber

18. Noises can be caused by worn suspension components, such as loose _____ _____ and strut _____ plates.

19. A _____ _____ is often necessary when trying to determine the cause of a noise or vibration.

20. Which of the following components has the primary purpose of reducing body sway?
    a. Shock absorber
    b. Control arm bushing
    c. Stabilizer bar
    d. Rebound bumper

21. You should also perform a search of the vehicle's service _____ and for any relevant technical service _____.

22. The most common suspension type in use today is the _____ suspension.

23. Label the components of the MacPherson strut shown in Figure 7-1.

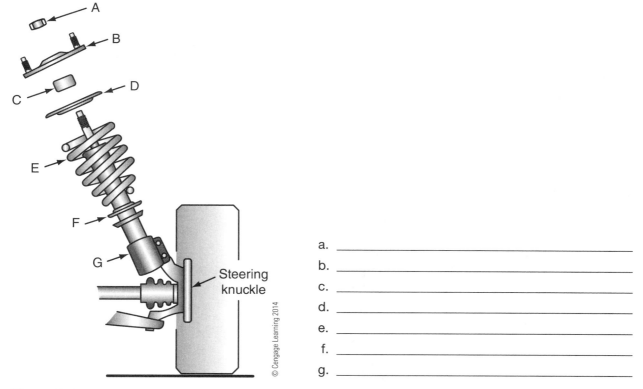

a. _____
b. _____
c. _____
d. _____
e. _____
f. _____
g. _____

Figure 7-1

24. Strut service usually requires _____ the strut from the vehicle.

25. Describe the process of disassembling and reassembling a MacPherson strut.

_____
_____
_____
_____

26. In some vehicles, the strut _____ can be replaced while the strut is still installed in the vehicle.

27. When you are replacing gas-charged struts or shocks, the gas charge needs to be _____ before throwing the strut or shock away.

28. When replacing shock absorbers, sometimes a _____ is needed to support the axle.

29. A bumper that appears _____ means that it has been contacting the frame, which can indicate a weak _____.

30. Which of the following is often necessary when replacing a steering knuckle?
    a. Removing the shock absorber
    b. Removing the tie rod
    c. Separating a ball joint
    d. All of the above

31. When replacing a steering knuckle: Technician A says the ball joint(s) must be torqued properly to support the vehicle's weight. Technician B says always install a new cotter pin when tightening a ball joint with a castle nut. Who is correct?
    a. Technician A
    b. Technician B
    c. Both A and B
    d. Neither A nor B

32. What is the purpose of the tool shown in Figure 7-2?

Figure 7-2

    a. Separating ball joints
    b. Installing ball joints
    c. Removing an axle
    d. Removing wheel studs

Chapter 7   Suspension System Service   153

33. Label the jacking and supporting locations to check ball joint wear in Figure 7-3.

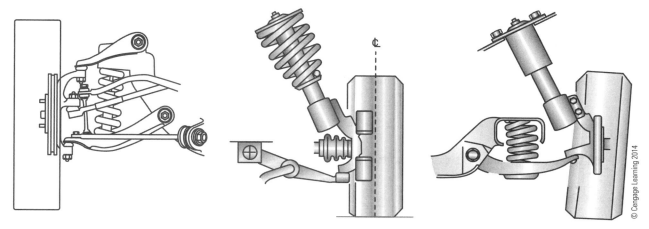

Figure 7-3

34. Explain the typical procedure to check ball joint wear for a load-carrying ball joint mounted in a lower control arm. _____
_____
_____

35. Some ball joints have built-in _____ _____ that change how far the grease fitting protrudes from the base of the joint.

36. To remove and install ball joints on many types of vehicles, a ball joint _____ is necessary.

37. On some vehicles it is necessary to replace the _____ _____ in order to replace the ball joint.

38. Control arm _____ replacement often requires special bushing tools.

39. Describe the problems caused by loose and worn radius arm bushings. _____
_____
_____

40. Explain the purpose of the sway bar in the front suspension. _____
_____

41. Grease or _____ fittings are used to lubricate suspension and steering joints.

42. Many rear struts bolt either into the _____ or into the rear firewall area where the rear seatback and parcel shelf are located.

Chapter 7 Suspension System Service

43. When you are replacing the rear shocks, the rear _____ should be supported.

44. A bent _____ bar will cause the rear axle to be misaligned to the body and will affect wheel alignment.

45. Explain why a complete prealignment inspection is important. _____
_____
_____

46. List five items that should be checked during a prealignment inspection.
    a. _____
    b. _____
    c. _____
    d. _____
    e. _____

47. Describe how and why to perform a dry-park check. _____
_____
_____
_____

Activities

1. Label the tools shown in Figure 7-4 (a through g).

Figure 7-4

2. For the suspension shown in Figure 7-5, identify the components which, if excessively worn, will adversely affect wheel alignment.

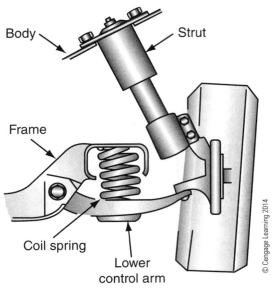

Figure 7-5

_____
_____

3. For the suspension shown in Figure 7-6, identify the components which, if excessively worn, will adversely affect wheel alignment.

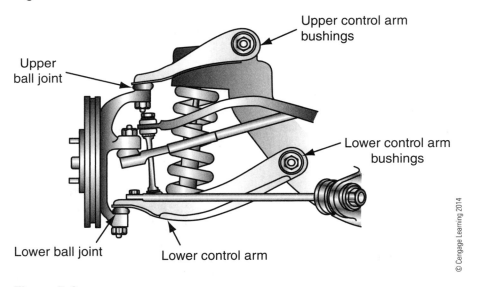

Figure 7-6

_____
_____

4. Match the following complaints with their most likely causes.

   Excessive body lean   Worn shocks
   Creaking noise   Worn control arm bushing
   Knocking over bumps   Worn ball joint
   Excessive bounce   Broken sway bar link
   Low ride height   Weak spring

Lab Worksheet 7-1

Name _____ Date _____ Instructor _____

Suspension Inspection

Date: _____

Name: _____

Address: _____

City: _____ State: _____ Zip Code: _____

Year: _____ Make: _____ Model: _____

Mileage: _____ License: _____ VIN: _____ Eng: _____

Customer Interview:
_____
_____
_____
_____
_____

PREALIGNMENT INSPECTION CHECKLIST

Owner _____ Phone _____ Date _____

Address _____ VIN _____

Make _____ Model _____ Year ____ Lic. number _____ Mileage _____

1. Road test results	Yes	No	Right	Left
Above 30 MPH				
Below 30 MPH				
Bump steer				
When bracking				
Steering wheel movement				
Stopping from 2–3 MPH (Front)				
Vehicle steers hard				
Strg wheel returnability normal				
Strg wheel position				
Vibration	Yes	No	Frnt	Rear

2. Tire pressure	Specs Frnt ____ Rear ____
Record pressure found	
RF ____ LF ____ RR ____ LR ____	

3. Chassis height	Specs Frnt ____ Rear ____
Record height found	
RF ____ LF ____ RR ____ LR ____	

	Yes	No
Springs sagged		
Torsion bars adjusted		

4. Rubber bushings	OK
Upper control arm	
Lower control arm	
Sway bar/stabilizer link	
Strut rod	
Rear bushing	

5. Shock absorbers/struts	Frnt	Rear

6. Steering linkage	Frnt OK	Rear OK
Tie-rod ends		
Idler arm		
Center link		
Sector shaft		
Pitman arm		
Gearbox/rack adjustment		
Gearbox/rack mounting		

7. Ball joints			OK
Load bearings			
	Specs	Readings	
	Right ____ Left ____	Right ____ Left ____	
Follower			
Upper strut bearing mount			
Rear			

8. Power steering	OK
Belt tension	
Fluid level	
Leaks/hose fittings	
Spool valve centered	

9. Tires/wheels	OK
Wheel runout	
Condition	
Equal tread depth	
Wheel bearing	

10. Brakes operating properly

11. Alignment	Spec		Initial reading		Adjusted reading	
	R	L	R	L	R	L
Camber						
Caster						
Toe						

Bump steer	Toe change right wheel		Toe change left wheel	
	Amount	Direction	Amount	Direction
Chassis down 3"				
Chassis up 3"				

	Spec		Initial reading		Adjusted reading	
	R	L	R	L	R	L
Toe-out on turns						
SAI						
Rear camber						
Rear total toe						
Rear indiv. toe						
Wheel balance						
Radial tire pull						

INSTRUCTOR VERIFICATION:

Lab Worksheet 7-2

Name _____ Date _____ Instructor _____

Year _____ Make _____ Model _____

Identify Causes of Tire Wear

1. Inspect the front and rear tires for abnormal wear patterns.

Conditions	Rapid wear at shoulders	Rapid wear at center	Cracked treads	Wear on one edge	Feathered edge	Diagonal wipe rear tire FWD vehicles	Scalloped wear
Effect							
Causes	Underinflation or lack of rotation	Overinflation or lack of rotation	Underinflation or excessive speed	Excessive camber	Incorrect toe	Incorrect wheel toe	Lack of rotation of tires or worn or out-of-alignment suspension
Corrections	Adjust pressure to specifications when tires are cool. Rotate tires.			Adjust camber to specs	Adjust toe to specs	Perform rear wheel alignment	Rotate tires and inspect suspension

LF _____

RF _____

LR _____

RR _____

2. Explain the possible causes for the wear patterns on the tires. _____

_____

_____

3. What suspension components can cause the types of tire wear shown? _____
   _____
   _____

4. How can the components listed above be inspected for wear? _____
   _____
   _____

5. Based on your findings, what do you recommend as the next steps to correct the problem? _____
   _____
   _____

INSTRUCTOR VERIFICATION:

Lab Worksheet 7-3

Name _____ Date _____ Instructor _____

Year _____ Make _____ Model _____

Front suspension type _____ Rear suspension type _____

Shock/Strut Inspection

Inspect the shocks and struts for oil leaks, noise, mounting hardware, and excessive bounce.

1. Visual inspection
   RF _____
   LF _____
   RR _____
   LR _____

2. Bounce each corner of the vehicle while listening for noises such as creaking and knocking. Note your findings. _____
   _____

3. If noise is heard, how can the shock/strut be isolated to determine if it is the cause of the noise? _____
   _____

4. Bounce each corner of the vehicle four or five times and count the number of times the vehicle bounces until it stops.
   RF _____
   LF _____
   RR _____
   LR _____
   Based on this test, what condition are the shock/struts? _____
   _____
   _____

5. Inspect the shock mountings and bushings for signs of wear and note your findings. _____
   _____
   _____

6. Based on your inspection of the shocks/struts, what are your recommendations? _____
   _____
   _____

INSTRUCTOR VERIFICATION:

Lab Worksheet 7-4

Name _____ Date _____ Instructor _____

Year _____ Make _____ Model _____

Front suspension type _____ Number of ball joints _____

Ball Joint Inspection

1. Locate and record the vehicle manufacturer's procedure for checking ball joint wear. _____
   _____

   Maximum allowed movement _____
   Rotation torque _____

2. Raise and support the vehicle as indicated in the ball joint inspection procedure. Note lifting locations.
   _____
   _____

3. Rock the tire in and out at the top and bottom and check for looseness. Note your findings. _____
   _____

4. Using a pry bar, pry up on the bottom of the tire and watch the ball joint for movement. Is there movement? _____ Yes   _____ No

   If yes, install a dial indicator to measure the amount of movement and note your results.
   Indicated movement _____
   Does this exceed the manufacturer's spec?   _____ Yes   _____ No

5. Ball joints that have a rotational torque spec are tested by using a beam or dial torque wrench to measure the force required to rotate the ball joint socket. This test is performed with the ball joint disconnected from the steering knuckle. Install the ball joint nut onto the stud and turn the joint with the torque wrench. Note the torque. _____
   _____

INSTRUCTOR VERIFICATION:

Lab Worksheet 7-5

Name _____ Date _____ Instructor _____

Year _____ Make _____ Model _____

Front suspension type _____ Number of ball joints _____

Electronic Suspension Inspection

1. Turn the ignition on and note the dash warning light for the electronic suspension system: does the light illuminate during bulb check? _____ Yes _____ No

2. Start the engine. Does the light go out after several seconds? _____ Yes _____ No

3. What is indicated if the light remains on? _____

4. With the vehicle raised, determine what suspension components are electronically controlled.
   Shocks _____ Struts _____ Sway bar/links _____
   Other _____
   How is/are the components above monitored and/or controlled? _____
   _____

5. Inspect the components for problems such as loose or disconnected wiring, broken air lines, air leaks, and oil leaks from shocks/struts. Note your findings. _____

6. Locate and record the procedure to retrieve trouble codes and data from the electronic suspension system.
   _____

7. Determine if there are any special service procedures or tools required to service components of the electronic suspension system. _____

8. Check and record any TSBs related to the electronic suspension system. _____
   _____
   _____

INSTRUCTOR VERIFICATION: _____

CHAPTER 8
Steering System Principles

Review Questions

1. The _____ system works with components of the suspension to provide for the turning movement of the wheels.

2. List three functions of the steering system.
   a. _____
   b. _____
   c. _____

3. The steering column enables the driver to control the direction of the vehicle and provides some _____ to make steering a little easier.

4. Modern vehicles have _____ steering columns to help prevent injury in front-end collisions.

5. _____ steering systems require most of the effort needed to turn the wheels to be supplied by the driver.

6. Hydraulic power steering uses a _____ _____ hydraulic power steering pump.

7. Modern power steering systems use which of the following to supply power assist?
   a. Electricity
   b. Hydraulics
   c. Pneumatics
   d. Both a and b

8. Leverage, also called _____ _____, is used at the steering wheel and the gearbox to increase force supplied by the driver.

9. The gearbox uses two gears to convert _____ motion of the steering wheel into a _____ motion that moves the wheels.

169

170   Chapter 8 Steering System Principles

10. Label the components of the recirculating ball gearbox shown in Figure 8-1 (a through i).

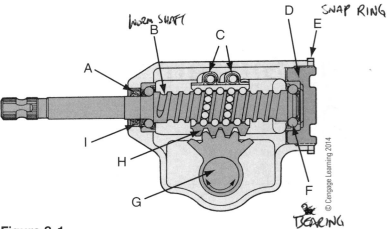

Figure 8-1

B — worm shaft
D — snap ring
F — bearing

11. Describe the operation of a recirculating ball gearbox. _____
_____
_____
_____

12. Define steering ratio. _____
_____
_____

13. Technician A says all power steering systems use the same power steering fluid. Technician B says automatic transmission fluid may be used in place of power steering fluid in modern vehicles. Who is correct?
    a. Technician A
    b. Technician B
    c. Both A and B
    d. Neither A nor B

14. Identify the components of the rack-and-pinion gearbox shown by unlabeled arrows in Figure 8-2.

15. Benefits of the rack-and-pinion gearbox include the reduced _____ and the elimination of several pieces of steering _____.

16. Explain how tires affect steering and how the vehicle handles. _____
_____
_____

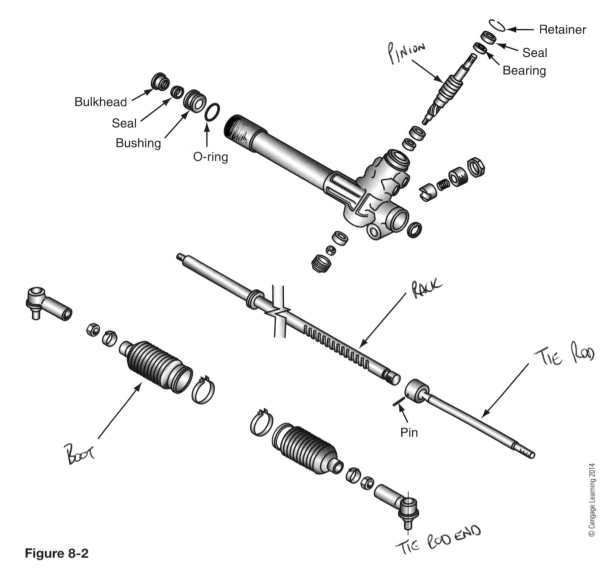

Figure 8-2

17. List five functions of the steering column in addition to its purpose for steering the vehicle. _____
_____
_____

18. Technician A says all collapsible steering columns use the plastic mesh construction that allows the column to collapse. Technician B says collapsible steering columns are important safety features in modern cars and trucks. Who is correct?
    a. Technician A
    b. Technician B
    c. Both A and B
    d. Neither A nor B

19. Another method used to prevent injury from the steering assembly during a collision is using a _____ steering shaft.

20. Which of the following components maintains the airbag connection through the steering column?
    a. Impact sensor
    b. Clock spring
    c. Airbag coupler
    d. None of the above

21. The _____ function allows the driver to adjust steering wheel position to increase comfort while driving.

22. A _____ steering wheel can move closer or farther from the driver's seat.

23. Describe the purpose of the ball bearings in the recirculating ball gearbox.
    _____
    _____

24. Label the power-assist components of the recirculating ball gearbox shown in Figure 8-3.

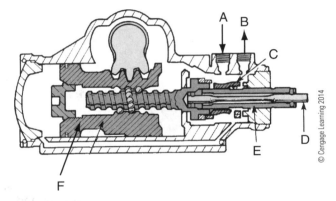

Figure 8-3

25. Label the parts of the parallelogram linkage shown in Figure 8-4.

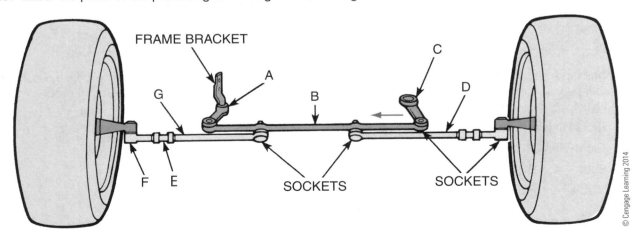

Figure 8-4

26. Describe the purpose of the idler arm. _____
    _____
    _____

27. In a parallelogram linkage, _____ assemblies connect the centerlink to the steering arm of the steering knuckle.

28. The tie rods are parallel with the lower control arms to prevent _____ steer.

29. Four-wheel-drive and I-beam-equipped vehicles often use a _____ linkage instead of a parallelogram linkage.

30. Most modern cars and light trucks use a _____ and _____ gearbox.

31. Technician A says both a recirculating ball and a rack-and-pinion gearbox use inner and outer tie rods. Technician B says rack-and-pinion gearboxes only use outer tie rod ends. Who is correct?
    a. Technician A
    b. Technician B
    c. Both A and B
    d. Neither A nor B

32. Describe how power assist is obtained through the pinion gear of a power-assisted rack-and-pinion gearbox.
    _____
    _____
    _____

33. Technician A says rack-and-pinion gearboxes can use either hydraulic or electric power assist. Technician B says electrically assisted rack-and-pinions can be used on both the front and rear wheels. Who is correct?
    a. Technician A
    b. Technician B
    c. Both A and B
    d. Neither A nor B

34. The check valve in a power steering pump is used to:
    a. limit maximum pressure.
    b. control fluid flow at idle.
    c. boost fluid pressure at high speed.
    d. bypass fluid flow at idle.

35. Explain the purpose of a power steering pressure switch. _____
    _____
    _____

Chapter 8 Steering System Principles

36. Electric power assist can be located:
    a. as part of the rack-and-pinion.
    b. as part of the ball nut.
    c. in the steering column.
    d. both a and c.

37. A hydraulic power-assist system uses a pump, a high-pressure hose, and a low-pressure _____ hose.

38. A worn and/or loose power steering belt can cause all of the following except:
    a. erratic power assist.
    b. excessive power assist.
    c. noise.
    d. increased steering effort

39. Explain two advantages of electric power steering assist. _____
_____
_____

Activities

1. Identify the steering components shown in Figure 8-5 (a through d).

a.

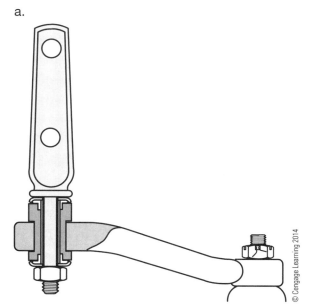

b.

c.

d.

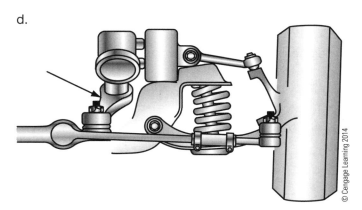

Figure 8-5

INSTRUCTOR VERIFICATION:

2. Label the components of the power steering system shown in Figure 8-6.

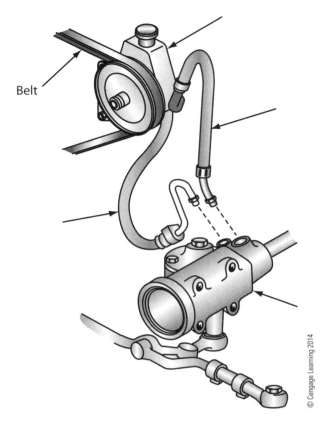

Belt

Figure 8-6

3. Calculate the steering ratios of the two examples shown in Figure 8-7 (a and b).

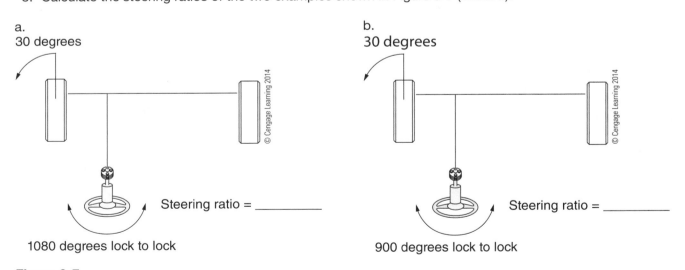

a.
30 degrees

Steering ratio = _____

1080 degrees lock to lock

b.
30 degrees

Steering ratio = _____

900 degrees lock to lock

Figure 8-7

a. Which of the two systems will provide easier steering? _____

b. Which of the two systems will have more road feel? _____

c. Explain why different steering ratios are used for different vehicles. _____

INSTRUCTOR VERIFICATION:

Lab Worksheet 8-1

Name _____ Date _____ Instructor _____

Year _____ Make _____ Model _____

Identify Steering Linkage Types and Components

1. Steering gearbox type _____

2. Type of steering linkage (circle one)   Parallelogram   Rack-and-pinion

    Haltenburger   Crosslink   Other

3. Components used (circle all that apply)   Inner tie rods   Outer tie rods

    Tie rod sleeves   Centerlink   Pitman arm   Idler arm   Drag link

    Year _____ Make _____ Model _____

1. Steering gearbox type _____

2. Type of steering linkage (circle one)   Parallelogram   Rack-and-pinion

    Haltenburger   Crosslink   Other

3. Components used (circle all that apply)   Inner tie rods   Outer tie rods

    Tie rod sleeves   Centerlink   Pitman arm   Idler arm   Drag link

    Year _____ Make _____ Model _____

1. Steering gearbox type

2. Type of steering linkage (circle one) Parallelogram   Rack-and-pinion

    Haltenburger   Crosslink   Other

3. Components used (circle all that apply)   Inner tie rods   Outer tie rods

    Tie rod sleeves   Centerlink   Pitman arm   Idler arm   Drag link

INSTRUCTOR VERIFICATION:

© 2014 Cengage Learning. All Rights Reserved. May not be scanned, copied or duplicated, or posted to a publicly accessible web site, in whole or in part.

Lab Worksheet 8-2

Name _____ Date _____ Instructor _____

Year _____ Make _____ Model _____

Steering Column Inspection

1. Steering column components

    Tilt _____ Telescoping _____

    Memory _____ Electric assist _____

2. Operate the tilt and telescoping functions (if applicable). Note any of the following.
    a. Binding, tightness, looseness _____
    b. Noises _____

3. With the engine off and the steering unlocked, turn the steering wheel from lock to lock. Note any noises from the column. _____

4. Start the engine and turn the wheel from lock to lock. Note any indication of binding or looseness in the column. _____

5. Based on your inspection, describe the condition of the steering column. _____

INSTRUCTOR VERIFICATION: _____

Lab Worksheet 8-3

Name _____ Date _____ Instructor _____

Year _____ Make _____ Model _____

Inspect Power Steering (Hydraulic Assist)

1. Describe the power steering assist system used on this vehicle. _____
   _____

2. List common causes of power steering fluid leaks. _____
   _____

3. Inspect the components from which fluid leaks often occur. Note the results of your inspection. _____
   _____

4. Locate the power steering fluid reservoir: inspect and note fluid level and appearance. _____

5. Remove a small sample of fluid from the reservoir and place the fluid on a clean, dry shop towel. Inspect the fluid for signs of metal. Note your findings. _____
   _____

6. Replace the reservoir cap. Inspect the pump drive belt and note its condition. _____
   _____

7. Start the engine and listen for noise from the power steering pump with the engine at idle and the wheels straight ahead. Note any noise from the pump. _____

8. Turn the wheels from lock to lock and note any noise from the pump. _____
   _____

9. With the help of an assistant turning the wheels, inspect the pump, lines, and gearbox for signs of fluid loss while turning the wheels from lock to lock. Note your findings. _____
   _____

10. Based on your inspection, summarize the condition of the power steering hydraulic system. _____
    _____

INSTRUCTOR VERIFICATION:

Lab Worksheet 8-4

Name _____ Date _____ Instructor _____

Year _____ Make _____ Model _____

Inspect Power Steering (Electric)

1. Describe the power steering assist system used on this vehicle. Note the location of the power-assist motor.
   _____

2. Turn the ignition on and note the electric power steering (EPS) warning lamp in the instrument cluster.
   Light on _____   Light off _____
   a. If there is no EPS light, determine if the vehicle manufacturer uses a warning indicator for a steering system fault.   Light used   Yes _____   No _____
   b. If a warning light is not used, how is a fault in the EPS system determined? _____
   _____

3. Check for EPS DTCs.   DTCs present   Yes _____   No _____
   If yes, record the DTCs _____

4. Connect a scan tool to the DLC. Note all EPS data PID values KOEO (key on engine off).

   PID_____   PID_____   PID_____
   PID_____   PID_____   PID_____
   PID_____   PID_____   PID_____
   PID_____   PID_____   PID_____

5. Start the engine and turn the steering from lock to lock. Does the steering operate smoothly?
   Yes _____   No _____

6. Summarize the results of your inspection of the EPS system. _____
   _____

INSTRUCTOR VERIFICATION:

CHAPTER 9
Steering Service

Review Questions

1. To remove a steering wheel, a steering wheel _____ kit is often needed.

2. To remove a _____ arm, a puller is required because of the tapered fit between the arm and the gearbox.

3. Follow the manufacturer's service procedures exactly when working on the _____ system.

4. Removing the airbag or SRS _____ is a common step when disarming the airbag system.

5. Explain why removing the battery cable is not always recommended as part of the airbag disarming process.
   _____
   _____
   _____

6. Spending time gathering _____ from the customer can often save time in the actual diagnosis.

7. Technician A says problems with the suspension system components cannot affect the steering system. Technician B says the two systems share some components and both systems should be inspected. Who is correct?
   a. Technician A
   b. Technician B
   c. Both A and B
   d. Neither A nor B

8. Describe how to check power steering fluid level and condition. _____
   _____
   _____
   _____

9. Describe how to remove an inner tie rod from an end-takeoff rack and pinion. _____
_____
_____
_____

10. The tool shown in Figure 9-1 is used to:
    a. remove the outer tie rod end.
    b. remove the rack-and-pinion boot.
    c. torque the castle nut.
    d. remove and install the inner rack tie rod.

Figure 9-1

11. When a tie rod has been replaced, the _____ angle setting should be checked and if necessary, adjusted to specs.

12. When installing a new component that uses a castle nut, Technician A tightens the nut to specs and then loosens it until the cotter pin hole aligns with the nut. Technician B tightens the nut to specs and then continues to tighten the nut until the cotter pin hole aligns. Who is correct?
    a. Technician A
    b. Technician B
    c. Both A and B
    d. Neither A nor B

13. Vehicles with recirculating ball gearboxes, such as those with SLA suspensions, usually use the _____ steering linkage.

14. Centerlinks and Pitman arms can be either wear or _____ types.

15. If nylon friction nuts are used instead of castle nuts, they should be _____ once removed.

16. Technician A says it is OK to reuse an old cotter pin if it is still in good condition. Technician B always replaces old cotter pins with new. Who is correct?
    a. Technician A
    b. Technician B
    c. Both A and B
    d. Neither A nor B

17. The tool shown in Figure 9-2 is used to:
    a. install tie rod ends.
    b. remove Pitman arms.
    c. install idler arms.
    d. none of the above.

Figure 9-2

18. Technician A says idler arms should have zero play in the sockets. Technician B says some idler arms are allowed a slight amount of play. Who is correct?
    a. Technician A
    b. Technician B
    c. Both A and B
    d. Neither A nor B

19. A steering damper is similar to a _____ _____ .

20. Which of the following is least likely to be the cause of a hard steering complaint?
    a. Defective power steering pump
    b. Worn power steering gearbox
    c. Worn bearings in the steering column
    d. Low power steering fluid level

21. Electric power assist can be applied either at the _____ _____ or to the rack and pinion.

22. Technician A says all modern vehicles can use either ATF or power steering fluid in the power steering system. Technician B says some vehicle manufacturers require special power steering fluid in their systems. Who is correct?

    a. Technician A
    b. Technician B
    c. Both A and B
    d. Neither A nor B

23. Explain what service is being performed in Figure 9-3. _____
    _____
    _____

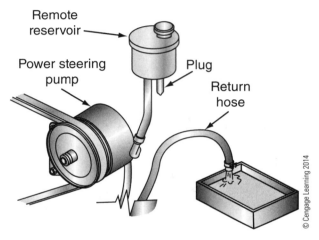

Figure 9-3

24. When flushing and bleeding a power steering system, why should the steering be held in the full-turn lock position for more than a few seconds? _____
    _____
    _____

25. Some power steering systems use a _____ inline with a power steering hose or in the reservoir.

26. Power steering hoses wear from the _____ out.

27. When a leaking hose is being replaced, it is a good idea to replace _____ hoses.

28. Describe how to replace a power steering hose. _____
    _____
    _____
    _____

29. Power steering belts should be _____ during routine service and often require _____ every 40,000 to 60,000 miles.

30. Describe three ways in which tension is applied to a power steering belt. _____
_____
_____

31. Poor power steering performance may be caused by all of the following except:
    a. loose power steering belt.
    b. worn belt tensioner.
    c. binding steering coupler.
    d. low power steering fluid level.

32. Some vehicles use a combination of _____ and _____ to provide variable power assist.

33. Some vehicles use electric _____ to provide all of the power steering assist.

34. Components of the _____-_____ system on a hybrid vehicle are bright orange.

35. Begin inspecting the electric power-assist system by noting if any _____ lights are illuminated on the dash.

36. A fault in the electric power assist may result in the system setting a _____ _____ code.

Activities

1. Make a list of steering and suspension components that can cause hard steering.

   _____
   _____
   _____
   _____
   _____

INSTRUCTOR VERIFICATION:

Lab Worksheet 9-1

Name _____ Date _____ Instructor _____

Year _____ Make _____ Model _____

Identify SRS Components

1. Determine the type(s) of air bags installed in the vehicle:

   Driver _____

   Passenger _____

   Side impact _____

   Curtain _____

   Knee _____

   Other _____

2. Does the vehicle have a passenger air bag on/off switch?   Yes _____   No _____

   If yes, describe the location _____

3. Does the vehicle have an occupant classification system?   Yes _____   No _____

   If yes, what types of sensors are used? _____

4. Using service information, determine the location of the sensors used by the SRS.

   Sensor type _____   Location _____
   Sensor type _____   Location _____
   Sensor type _____   Location _____
   Sensor type _____   Location _____

INSTRUCTOR VERIFICATION:

Lab Worksheet 9-2

Name _____ Date _____ Instructor _____

Year _____ Make _____ Model _____

Disable and Enable the SRS

1. Locate and record the manufacturer's service procedures for disarming the air bag system. _____
   _____
   _____

2. If a wait time is specified after disarming the system but before work begins, why is it important for you to follow this step? _____
   _____

3. If the service procedure includes disconnecting the battery, what other systems and components will be affected?
   _____
   _____

4. What items may need to be reset after the battery is disconnected and reconnected?
   _____
   _____

5. Perform the procedure to disarm the air bag.
   Instructor check _____

6. Re-enable the air bag.
   Instructor check _____

7. Turn the ignition on and note the SRS warning lamp; does the lamp illuminate?
   Yes _____   No _____

8. If yes, start the engine and note the lamp after several seconds. Did the lamp go out? _____
   Yes _____   No _____
   If yes, what does this indicate? _____

9. If the lamp did not illuminate in step 8, recheck your work to ensure the system has been properly re-enabled.
   Did you find and correct the problem?   Yes _____   No _____
   Instructor check _____

INSTRUCTOR VERIFICATION:

Lab Worksheet 9-3

Name _____ Date _____ Instructor _____

Year _____ Make _____ Model _____

Steering System Inspection

Steering gear type _____ Linkage arrangement _____

    Date: _____
    Name: _____
    Address: _____
    City: _____ State: _____ Zip Code: _____
    Year: _____ Make: _____ Model: _____
    Mileage: _____ License: _____ VIN: _____ Eng: ____

Customer Interview:
_____
_____
_____
_____
_____

INSTRUCTOR VERIFICATION:

Lab Worksheet 9-4

Name _____ Date _____ Instructor _____

Year _____ Make _____ Model _____

Replace a Power Steering Belt

1. Locate the power steering pump belt. Belt type _____

2. Describe how the belt is loosened or tension is removed. _____
   _____

3. What tools are required to remove the belt? _____
   _____

4. Do any other belts need to be removed to remove the power steering belt?
   Yes _____   No _____
   If yes, what belt? _____

5. Remove the power steering belt and note its condition. _____
   Instructor check _____

6. Locate the manufacturer's specs for proper belt tension and record. _____

7. Install the belt and apply tension. Check for proper tension with a gauge and record your findings. _____
   _____
   Instructor check _____

8. Make sure the belt is properly aligned and seated in all pulleys. Start the engine and confirm proper operation.
   Instructor check _____

INSTRUCTOR VERIFICATION:

Lab Worksheet 9-5

Name _____ Date _____ Instructor _____

Year _____ Make _____ Model _____

Inspect Power Steering Fluid

1. Locate and record the manufacturer's recommended power steering fluid. _____

2. Determine the location of the power steering fluid reservoir.
   Remote reservoir _____   Pump/reservoir _____

3. If the fluid level can be seen through the reservoir, note the level. _____

4. If the cap must be removed to check the level, clean the cap and reservoir around the cap before removing. Remove the cap and note the level on the dipstick. _____

5. Note the color of the fluid. _____

6. Remove a few drops of fluid and wipe on a clean, light-colored shop towel. Inspect the fluid for metal shavings, color, and smell. Note your findings. _____

7. Based on your inspection, what is the condition of the fluid? _____

INSTRUCTOR VERIFICATION:

Lab Worksheet 9-6

Name _____ Date _____ Instructor _____

Year _____ Make _____ Model _____

Inspect for Power Steering Fluid Loss

1. Check and record power steering fluid level. _____

2. Inspect the following locations and note any signs of fluid leakage.
    Power steering pump   _____
    Power steering pressure hose   _____
    Power steering return hose   _____
    Power steering pressure switch _____
    Pinion seals/worm shaft seals _____
    Rack fluid transfer lines   _____
    Rack lower pinion seals   _____
    Pitman shaft seals   _____
    Rack end seals (bellows)   _____
    Power steering cooler lines   _____
    Power steering cooler   _____

3. Based on your inspection, what is the necessary action? _____
_____
_____

INSTRUCTOR VERIFICATION:

Lab Worksheet 9-7

Name _____ Date _____ Instructor _____

Year _____ Make _____ Model _____

Inspect Electric Power Steering (EPS) Assist System

1. Turn the ignition on and note the EPS warning light on the dash. Does the light illuminate at key-on bulb check? Yes _____ No _____

2. Start the engine and note the EPS light. _____
   What does this indicate? _____

3. Connect a scan tool to the diagnostic link connector (DLC) and navigate to the EPS menu. Check for stored diagnostic trouble codes (DTCs). Note any stored or history DTCs. _____
   _____

4. Navigate to the EPS data menu. With the engine running and the steering wheel straight ahead, note the following data.

   Steering shaft torque   _____
   Steering motor temperature   _____
   Torque sensor volts/N·m   _____
   Steering angle sensor degrees   _____
   Motor amperage   _____

5. Navigate to the EPS data menu. With the engine running and the steering wheel turned 90 degrees and held, note the following data.

   Steering shaft torque   _____
   Steering motor temperature   _____
   Torque sensor volts/N·m   _____
   Steering angle sensor degrees   _____
   Motor amperage   _____

6. Based on your inspection, note any problems or concerns with the EPS. _____
   _____
   _____

INSTRUCTOR VERIFICATION:

CHAPTER 10
Brake System Principles

Review Questions

1. Modern brake systems combine the principles of _____, _____, and electronics.

2. Leverage is also called mechanical _____.

3. The brake pedal acts as a _____, increasing the amount of force applied to the master cylinder pushrod.

4. A typical brake pedal may have a leverage ratio of:
   a. 1:1.
   b. 3:1.
   c. 20:1.
   d. 100:1.

5. Leverage involves a trade-off: increasing _____ requires moving a greater distance.

6. The slight amount of pedal movement at the released position before the pushrod begins to move into the booster and master cylinder is called:
   a. pedal travel.
   b. pedal height.
   c. free play.
   d. compliance.

7. Technician A says the brake light switch may be used as an input for the on-board computer system. Technician B says some vehicles use vacuum-operated brake light switches. Who is correct?
   a. Technician A
   b. Technician B
   c. Both A and B
   d. Neither A nor B

207

8. Some vehicles have _____ pedals that can move based on driver preference.

9. _____ is the science of using liquids to perform work.

10. By pressurizing a fluid in a closed system, the fluid can transmit both _____ and _____.

11. Pascal determined that _____ exerted on a confined liquid caused an increase in pressure at all points.

12. Which of the following statements about hydraulics is incorrect?
    a. Hydraulic pressure in a closed system is constant.
    b. A smaller input piston generates less hydraulic pressure than a larger piston.
    c. If output pressure increases, output piston travel decreases.
    d. If output pressure decreases, output piston travel increases.

13. Describe in your own words how hydraulic principles are used in the hydraulic brake system. _____
    _____
    _____
    _____

14. The hydraulic brake system input cylinder is called the _____ _____.

15. If 500 pounds is applied to a master cylinder piston with two square inches of surface area, the pressure in the cylinder will be:
    a. 250 psi.
    b. 500 psi.
    c. 1,000 psi.
    d. 2,000 psi.

16. The front brake hydraulic output is called a disc brake _____.

17. Explain why the pistons used in the disc brake caliper are much larger than the pistons in the master cylinder. _____
    _____

18. Explain why the pistons used in rear drum brakes are much smaller than the pistons used in the front disc brakes. _____
    _____
    _____

19. To operate properly, the hydraulic brake system must be a _____ system, meaning there is no opening for fluid to leak or vent.

20. Why do modern master cylinders have two chambers and pistons? _____

21. Explain why some master cylinders use two different-sized pistons. _____

22. The reservoir cap seals are often _____ seals, meaning that they _____ in size as the fluid level in the reservoir drops.

23. Technician A says the master cylinder primary piston is located in the front of the master cylinder. Technician B says the master cylinder primary piston is located in the rear of the cylinder. Who is correct?
    a. Technician A
    b. Technician B
    c. Both A and B
    d. Neither A nor B

24. Which is the function of the proportioning valve?
    a. Limit pressure to the front disc brakes
    b. Limit pressure to the rear disc brakes
    c. Limit pressure to the rear drum brakes
    d. Illuminate the brake warning light on the dash

25. What is the function of the load-sensing proportioning valve?
    a. Limit pressure to the front disc brakes
    b. Control rear drum brake pressure based on vehicle load
    c. Shut off a brake circuit in the event of a leak
    d. Illuminate the brake warning light on the dash

26. The _____ valve is used to delay slightly the application of the front disc brakes.

27. Which valve is responsible for closing off one of the hydraulic circuits if a leak occurs?
    a. Metering valve
    b. Proportioning valve
    c. Pressure differential valve
    d. None of the above

Chapter 10 Brake System Principles

28. The pressure differential valve usually contains an _____ contact that completes the brake warning lamp circuit.

29. Instead of a vehicle having three separate brake valves, they are commonly all together in a _____ valve.

30. All of the following describe brake line material except:
    a. Flared ends to prevent leaks
    b. Made from corrosion resistant aluminum
    c. Has double wall thickness for strength
    d. Metric and standard flares not interchangeable

31. A _____ splits a single brake line so that each rear wheel brake receives brake fluid.

32. Calipers and wheel cylinders are the _____ of the hydraulic system.

33. A fixed caliper has at least how many pistons?
    a. Two
    b. Three
    c. Four
    d. Six

34. Which caliper component is responsible for retracting the piston when the brakes are released?
    a. Caliper return spring
    b. Caliper dust boot
    c. Square seal
    d. Caliper bracket guide

35. _____ calipers are the most common type of brake caliper used in modern cars and light trucks.

36. Newton's Third Law of Motion states that for every _____ there is an _____ and _____ reaction.

37. Explain how Newton's Third Law of Motion applies to the disc brake system. _____
_____
_____

38. List the components of a typical front brake caliper. _____
_____
_____

39. Explain the operation of the integral parking brake rear disc brake caliper. _____
_____
_____

40. Some newer vehicles use _____ operated parking brake calipers.

41. Wheel cylinders are used with the _____ brake arrangement.

42. Explain what DOT 3, DOT 4, and DOT 5 brake fluid ratings mean. _____
_____
_____

43. List five qualities important for brake fluid. _____
_____
_____

44. Technician A says DOT 3 and DOT 4 are compatible fluids. Technician B says DOT 4 should not be used with vehicles equipped with antilock brakes (ABS). Who is correct?
    a. Technician A
    b. Technician B
    c. Both A and B
    d. Neither A nor B

45. If a fluid is _____, that means that it easily absorbs moisture.

46. Why must a technician be careful when working with brake fluid? _____
_____
_____

47. The _____ of _____ is a number that expresses the ratio of force required to move an object divided by the mass of the object.

48. Define brake fade and describe how it can occur. _____
_____
_____
_____

49. Explain why heat dissipation is important for the brake system. _____
_____
_____

50. Technician A says solid brake rotors are able to dissipate heat more effectively than vented rotors. Technician B says vented brake rotors are standard equipment on the front of all modern cars and trucks. Who is correct?
    a. Technician A
    b. Technician B
    c. Both A and B
    d. Neither A nor B

51. In conventional vehicles when the brakes are applied, the _____ energy of the vehicle is converted into _____ energy at the brakes.

52. Hybrid electric vehicles use the _____ system to recover brake energy to _____ the high-voltage batteries.

INSTRUCTOR VERIFICATION:

Chapter 10 Brake System Principles

Activities

I. Leverage

The brake pedal in a car operates as a second-class lever. With a second-class lever, the fulcrum is at one end of the lever instead of the middle. The effort is applied to the other end of lever, and the force is applied somewhere in between the effort and the fulcrum, like that shown in Figure 10-1. Look under the dashboard of several vehicles to see how the brake pedal is mounted. Notice that the top of the pedal is the mounting point, which acts as the fulcrum. Below the fulcrum, the pedal pushrod is mounted and extends forward through the firewall, to the brake booster, and/or master cylinder. The footpad at the bottom of the pedal is where the effort is applied.

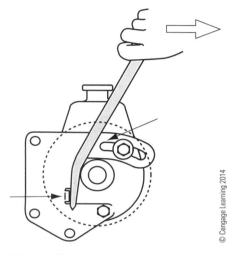

Figure 10-1

Measure the length of the pedal between the pushrod and fulcrum and the pushrod and the lowest point of the pedal for several vehicles.

1. Pushrod-to-fulcrum length _____ Pushrod-to-pedal length _____

2. Pushrod-to-fulcrum length _____ Pushrod-to-pedal length _____

Since the force is between the fulcrum and the effort, the resulting force is going to be determined by the distance below the force divided by the distance above the force. Use the following formula: $R = d_a/d_b$. R is the ratio, d_a is the distance from the pushrod to the end of pedal, and d_b is the distance from the pushrod to the fulcrum. Using the measurements of the brake pedals from questions 1 and 2, determine the ratio of force for each vehicle.

3. $d_a/d_b = $ _____ Ratio _____

4. $d_a/d_b = $ _____ Ratio _____

As you can see, the brake pedal and its mounting design reduce the amount of effort needed from the driver. This reduction of effort is beneficial for the driver, as it will reduce fatigue over the time of vehicle operation.

INSTRUCTOR VERIFICATION:

II. Hydraulics

Even though the force applied to the brake pedal is increased, it is necessary for the hydraulic system to further increase the force and apply it in a manner that will safely slow and stop the vehicle.

A simple hydraulic system, like that in Figure 10-2, contains two equal-sized containers of a liquid with two equal-sized pistons. The two containers are connected with a hose or tubing. If one piston is pushed downward with 100 lb (45 kg) of force and moves down one inch, the piston in the second container will move up one inch with the same 100 lb (45 kg) of force. Since the pistons are the same size, any force and movement imparted on one piston will cause the same reaction to the second piston.

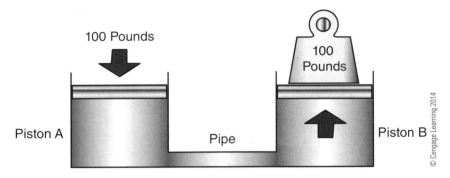

Figure 10-2

The use of hydraulics really provides advantage when the sizes of the pistons are different: the resulting force and movement can be increased or decreased as needed. The pressure generated by the piston is a factor of the piston size. Input pressure is found by dividing force by piston size, or $P = F/A$. The smaller the input piston surface area, the larger the force will be from that piston. The larger the input piston surface area, the less the force will be from that piston. Conversely, the output piston force is proportional to the pressure against the surface area of the piston. An output piston that is larger than the input piston will move with greater force than the input piston but over a shorter distance. An example is shown in Figure 10-3. To examine this principle we will look at what is known as Pascal's principle or law.

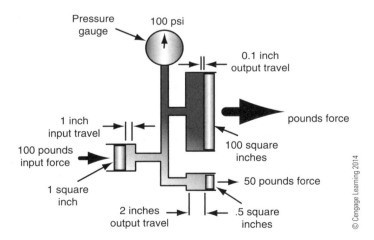

Figure 10-3

If the smaller piston on the left is our input piston and has a surface area of 1 square inch (6.45 cm^2) and the larger piston is our output piston and has an area of 10 square inches (65.5 cm^2), we can calculate the forces and movements quite easily. Suppose a force of 100 lb (45 kg) is exerted on our input piston, moving it 1 inch

(2.54 cm). Using $P = F/A$, we can calculate $P = 100/1$, or 100 psi. Our output piston, which has a surface area of 10 square inches, will receive 100 pounds of pressure per square inch. This will result in an output force by piston 2 of 1,000 lbs. However, since our output force has increased, our output movement will decrease. The larger piston will move one-tenth of the distance of the input piston; since the force was multiplied by 10, the distance will be divided by 10 as well. The 1 inch of movement of piston 1 turns into 1/10 of an inch of movement at piston 2. To calculate the forces and movement in a hydraulic circuit, use the following formulas:

$P_1 = F_1/A_1$ for piston 1 pressure
$F_2 = A_2/A_1 \cdot F_1$ for piston 2 force
$P_2 = F_2/A_2$ for piston 2 pressure

To practice using these principles, complete the following activity.

1. Calculate the forces and movements of a two-piston hydraulic circuit with an input piston of 2 square inches and an output piston of 10 square inches.

    Input distance _____ Input force _____
    Output distance _____ Output force _____

2. Calculate the forces and movements of a two-piston hydraulic circuit with an input piston of 2 square inches and an output piston of 1 square inch.

    Input distance _____ Input force _____
    Output distance _____ Output force _____

The pushrod from the brake pedal pushes on pistons inside the brake master cylinder. The master cylinder provides the pressure in the brake system. The force applied to the brake pedal, the ratio of brake pedal force multiplication, and the master cylinder piston size all determine how much pressure will be developed in the brake system. The size of the output pistons at the calipers and wheel cylinders, along with the brake system pressure, determine the final output force for the brakes. Use Figure 10-4 to determine brake system pressure.

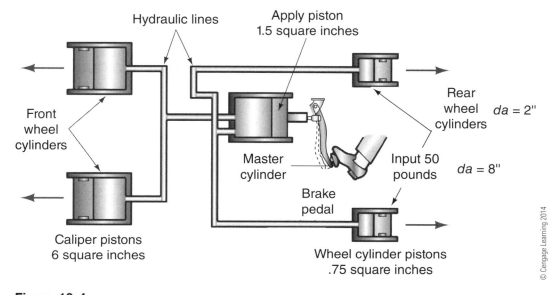

Figure 10-4

INSTRUCTOR VERIFICATION:

3. Input force to pedal _____

4. Brake pedal ratio D_a _____ D_b _____ Ratio _____

5. Hydraulic system pressure at master cylinder _____

6. Output force of caliper pistons _____

7. Output force of wheel cylinder pistons _____

III. Friction

Once the hydraulic system generates the pressure to actuate the pistons in the calipers and wheel cylinders, it is the job of the lining surfaces to slow the vehicle. The lining surfaces are comprised of the brake rotor or disc, brake pads, brake drum, and brake shoes. Brake pads are squeezed together against the rotor by the brake caliper. Brake fluid, under pressure from the hydraulic system, pushes on the caliper piston. As the piston moves out, pushing the inner pad against the rotor, the caliper slides inward, pressing the outer pad against the outside of the rotor. On drum brakes, the fluid pushes on two equal-diameter pistons in the wheel cylinder. The pistons move out, pressing the brake shoes against the inside surface of the brake drum.

Friction is created when uneven surfaces, in contact with each other, start to move relative to each other. Even though the two surfaces may appear smooth, there are slight imperfections that resist moving against each other. The amount of force required to move one object along another is called the coefficient of friction (CoF). Simply put, the CoF is equal to the ratio of force to move an object divided by the weight of the object, $CoF = F/M$. While several factors affect the CoF, such as temperature and speed, we will use a simple example of two bodies moving against each other. Imagine you have a 100 lb (45 kg) block of rubber on the floor of your lab. The force required to slide the block of rubber over concrete would be great. If it takes 100 lb (45 kg) of force to slide the block, the CoF will be 100/100 or 1.0. Imagine the same block of rubber now sitting on the floor of an ice hockey rink. If it only takes 15 lb (6.8 kg) of force to slide the block, what will the CoF be?

1. 15/100 = _____

To experiment with the CoF of various objects in your lab, you can make a small force gauge using an ordinary ballpoint pen, a rubber band, tape, and a paper clip. Assemble the parts as shown in Figure 10-5. Attach the paper clip to an object and try to pull it across a flat surface using the opposite end of your force meter. Measure with a ruler how much the pen tube extends from the body of the pen. This will give you an idea of how much force is being required to drag each object. Record your results below:

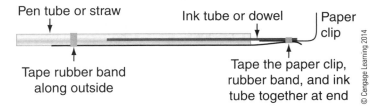

Figure 10-5

2. Object _____ Surface _____ Length of extension _____

3. Object _____ Surface _____ Length of extension _____

4. Object _____ Surface _____ Length of extension _____

5. Why did some objects require more force than others did? _____
   _____

6. How does the surface used to slide across affect the CoF? _____
   _____

7. If a liquid is placed between the two objects, what effect will that have on the CoF?
   _____

Static friction occurs between two stationary objects, and dynamic friction occurs between a stationary and a moving object. For example, an applied parking brake has static friction between the brake pad or shoe and the rotor or drum. When driving, there is dynamic friction between the brake pads and rotor when the driver applies the brakes.

8. List several examples of each type of friction related to the automobile.
   _____
   _____
   _____

9. What could happen if the CoF of the brake pads or shoes was too high? _____
   _____

10. What could happen if the CoF of the brake pads or shoes was too low? _____
    _____

11. What factors do you think are involved in determining the correct CoF for a particular vehicle? _____
    _____

The CoF of brake linings has a large impact on how well a vehicle will stop. Using lining materials with too high or low a CoF can cause brake performance issues, rapid wear, and customer dissatisfaction. Similarly, the tires, since they are the contact between the vehicle and the road, also play an important role in brake operation. Even the best-performing brakes will not stop a vehicle well if the tires cannot maintain proper traction.

INSTRUCTOR VERIFICATION:

IV. Brake Component Identification

1. How is the disc brake caliper in Figure 10-6a different than the caliper in Figure 10-6b?

a.

b.

Figure 10-6

2. Identify the parts of the master cylinder shown in Figure 10-7.

   a. _____
   b. _____
   c. _____
   d. _____
   e. _____

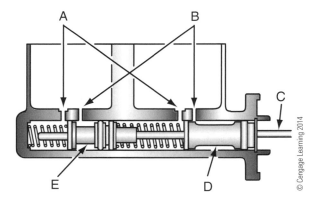

Figure 10-7

INSTRUCTOR VERIFICATION:

3. Identify the parts of the brake caliper shown in Figure 10-8.
   a. _____
   b. _____
   c. _____
   d. _____
   e. _____
   f. _____

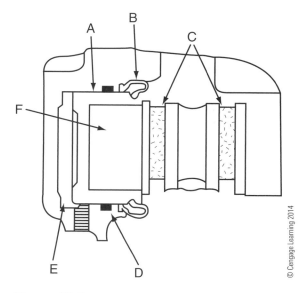

Figure 10-8

4. Identify the parts of the wheel cylinder shown in Figure 10-9.
   a. _____
   b. _____
   c. _____
   d. _____
   e. _____
   f. _____

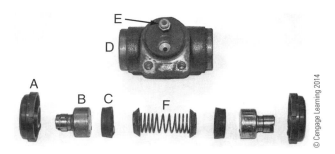

Figure 10-9

INSTRUCTOR VERIFICATION:

220 Chapter 10 Brake System Principles

5. Label the brake valves shown in Figure 10-10.

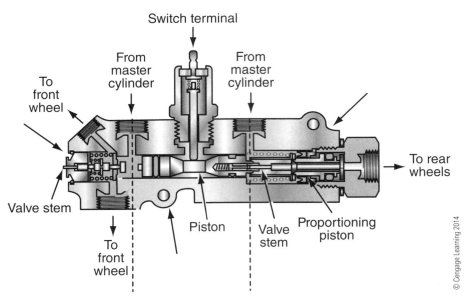

Figure 10-10

6. List examples of the different types of substances brake fluid is in contact with in the brake system.
___
___
___

7. How will a leak in the hydraulic system affect brake operation? Describe how the brake pedal may feel, how stopping distances may be affected, and what the customer complaint may include. ___
___
___
___
___

INSTRUCTOR VERIFICATION:

Lab Worksheet 10-1

Name _____ Date _____ Instructor _____

Year _____ Make _____ Model _____

Brake Pedal Leverage

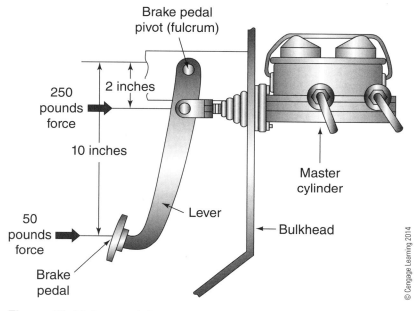

Figure 10-11 (example)

1. Using a tape measure, determine the total length of the brake pedal from the pad to the pivot. Record your measurement _____

2. Measure the distance between the pivot and the pushrod; record your measurement. _____

3. Determine the ratio of mechanical advantage gained by the brake pedal. _____

4. If the driver applies 100 pounds of force against the brake pedal, how much force is applied to the brake pushrod? _____

5. Describe what you think the pedal ratio may be if the vehicle were not equipped with power brakes.
   _____
   _____

INSTRUCTOR VERIFICATION:

6. Describe how you think the brake pedal assembly may be different on smaller and on larger vehicles than this vehicle. _____
_____
_____
_____

INSTRUCTOR VERIFICATION:

Lab Worksheet 10-2

Name _____ Date _____ Instructor _____

Hydraulics

Using a selection of brake hydraulic system components provided by your instructor, determine the amount of force generated by the hydraulic system using the Figure 10-12 and the piston sizes.

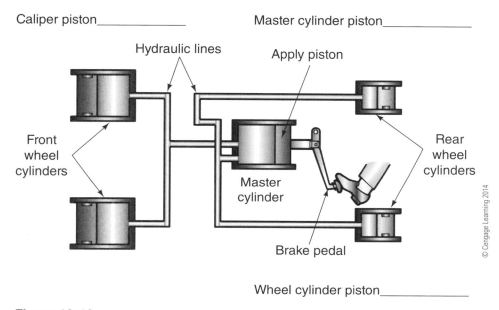

Figure 10-12

1. Master cylinder piston diameter _____

   Master cylinder piston area _____

2. Caliper piston diameter _____

   Caliper piston area _____

3. Wheel cylinder piston diameter _____

   Wheel cylinder piston area _____

INSTRUCTOR VERIFICATION:

Chapter 10 Brake System Principles

4. Using an input force of 50 pounds to the master cylinder, what is the pressure produced in the master cylinder? _____ Output force of the caliper _____ Output force of the wheel cylinder _____

5. If the master cylinder piston size is reduced, what happens to the pressure generated by the master cylinder? _____

_____

_____

INSTRUCTOR VERIFICATION:

Lab Worksheet 10-3

Name _____ Date _____ Instructor _____

Master Cylinders

1. Using a selection of master cylinders provided by your instructor, measure the diameter of the primary and secondary pistons.

   Cyl 1 _____ primary   _____ secondary

   Cyl 2 _____ primary   _____ secondary

   Cyl 3 _____ primary   _____ secondary

2. Based on your measurements, determine the surface area of the pistons. Use either πr^2 or $\text{diameter}^2 \times .785$.

   Cyl 1 _____ primary   _____ secondary

   Cyl 2 _____ primary   _____ secondary

   Cyl 3 _____ primary   _____ secondary

3. Based on your measurements, how much pressure will be developed by each master cylinder if the input force is 200 pounds?

   Cyl 1 _____ primary psi   _____ secondary psi

   Cyl 2 _____ primary psi   _____ secondary psi

   Cyl 3 _____ primary psi   _____ secondary psi

4. Based on your measurements, how does the pressure developed by the piston vary with different-sized pistons? _____
   _____
   _____

INSTRUCTOR VERIFICATION: _____

CHAPTER 11

Brake System Service

Review Questions

1. Line or _____ nut wrenches are used on brake line fittings.

2. A _____ tool is used to make new flares on a brake line.

3. A _____ bleeder applies pressure to the brake fluid at the master cylinder and forces air and fluid out of the hydraulic system.

4. Do not use _____-end wrenches to break loose tight fasteners.

5. Do not allow brake fluid to remain in contact with _____ surfaces, _____, plastics, or any other parts of the vehicle.

6. As a technician, you will be required to _____-_____ vehicles to verify customer concerns and to verify repairs have been completed properly.

7. Describe four problems you are checking for during a test drive.
   a. _____
   b. _____
   c. _____
   d. _____

8. The brake pedal assembly may be inspected for pedal _____, travel, and _____ play.

9. Incorrect brake pedal height may be caused by which of the following concerns?
   a. Low brake fluid level
   b. Improper parking brake adjustment
   c. Excessive shoe-to-drum clearance
   d. Faulty proportioning valve

228 Chapter 11 Brake System Service

10. Brake pedal _____ _____ is the very slight movement of the pedal before the brake pushrod begins to move.

11. List systems that may use the brake light switch as an input. _____

12. Explain how to test brake light switch operation. _____

13. Referring to Figure 11-1: Technician A says the adjustment shown is used to adjust the stoplight switch. Technician B says the adjustment shown is to preset the pushrod into the power brake booster. Who is correct?
    a. Technician A
    b. Technician B
    c. Both A and B
    d. Neither A nor B

Figure 11-1

14. Explain why the brake light circuit is often a hot-at-all-times circuit. _____

15. To test for a concern with power-adjustable pedals, a _____ _____ is needed to access DTCs and data.

16. When it is necessary to add brake fluid to the reservoir, always use _____ brake fluid from a _____ container.

17. Describe two causes for the brake fluid level to be low. _____
_____
_____

18. Explain why many vehicle manufacturers recommend periodic brake fluid changes.
_____
_____
_____

19. Which of the following does not cause the red brake warning light to illuminate?
    a. Low brake fluid level
    b. Contaminated brake fluid
    c. Loss of hydraulic pressure
    d. Activated parking brake

20. A late model vehicle with about 35,000 miles has low brake fluid level and the red BRAKE light is illuminated on the dash. Explain what can cause this and what should be recommended to the customer._____
_____
_____
_____

21. The _____ - _____ switch will turn on the red BRAKE light if there is an imbalance in pressure in the brake system.

22. Explain what can result if a petroleum-based fluid is added to the brake fluid.
_____
_____
_____

23. Explain what happens to brake fluid when it has been left in service for too long. _____
_____
_____
_____

24. A faulty flexible brake hose can cause which of the following brake concerns?
    a. Sticking brake caliper
    b. Non-functioning brake caliper
    c. Fluid loss
    d. All of the above

25. Over time, moisture absorbed by the brake fluid can cause the wheel cylinder pistons to _____ into their bores.

26. External fluid leaks from the master cylinder result from a leaking seal on the _____ piston.

27. Technician A says a leaking master cylinder primary piston seal will always result in fluid showing on the outside of the power brake booster. Technician B says some master cylinders can leak into the power brake booster assembly. Who is correct?
    a. Technician A
    b. Technician B
    c. Both A and B
    d. Neither A nor B

28. Describe the steps to replace a master cylinder. _____

29. Explain the steps to replace a rusted and leaking brake line. _____

30. Label the type of brake line flare Figure 11-2.

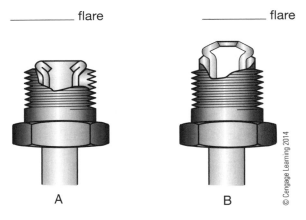

_____ flare   _____ flare

A   B

Figure 11-2

31. A vehicle's brake pedal sinks to the floor under normal pressure. Upon inspection, fluid level is correct and no external leaks are discovered. Which of the following is the most likely cause?
    a. Worn brake pads and rotors
    b. Kinked brake line

c. Bypassing master cylinder

d. None of the above

32. Which of the following is not a benefit of periodic brake fluid flushing?

    a. Decreased rust formation in hydraulic system

    b. Prolong service life of hydraulic components

    c. Reduced brake fluid boiling temperature

    d. Reduced risk of brake fade

33. Explain in detail three methods of brake system bleeding. _____
_____
_____
_____
_____

34. To properly bleed the ABS, a _____ _____ may be required.

35. Technician A says the tool shown in Figure 11-3 is required to bleed ABS-equipped vehicles. Technician B says the tool shown in Figure 11-3 is used to bleed the master cylinder. Who is correct?

    a. Technician A

    b. Technician B

    c. Both A and B

    d. Neither A nor B

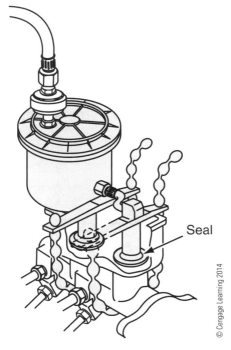

Figure 11-3

Activities

1. Label the brake tools shown in Figure 11-4 and Figure 11-5.

Figure 11-4

Figure 11-5

2. Describe the purpose and operation of the pressure differential valve. _____
_____
_____
_____

3. Describe the purpose and operation of the proportioning valve. _____
_____
_____
_____

4. Describe the purpose and operation of the load-sensing proportioning valve. _____
   _____
   _____

5. Explain the purpose and operation of the component shown in Figure 11-6. _____
   _____
   _____

Figure 11-6

INSTRUCTOR VERIFICATION:

Lab Worksheet 11-1

Name _____ Date _____ Instructor _____

Year _____ Make _____ Model _____

Brake Pedal Inspection

1. Locate and record the following brake pedal specs.
   Pedal free play _____
   Pedal height _____
   Pedal travel _____
   a. Why is brake pedal free play important for proper brake system operation?
   _____
   _____

   b. What other components can affect brake pedal height? _____
   _____

2. Operate the brake pedal and listen for noise. Note your findings. _____
   _____

3. Check and record the brake pedal free play. _____
   a. Is the free play within specs?   Yes _____   No _____
      If no, how can free play be adjusted? _____
      _____

4. Check and record pedal height. _____
   a. Is pedal height within specs?   Yes _____   No _____
   If no, what affects pedal height? _____
   _____

5. Check and record pedal travel. _____
   a. Is pedal travel within specs?   Yes _____   No _____
      If no, what affects pedal travel? _____
      _____

6. Based on your inspections, what actions are required? _____
   _____

INSTRUCTOR VERIFICATION:

Lab Worksheet 11-2

Name _____ Date _____ Instructor _____

Year _____ Make _____ Model _____

Brake Light Inspection

1. Locate the brake light switch and note its location. _____

2. Have an assistant press the brake pedal and note the operation of the brake lights.

   Are all lights operating?   Yes _____   No _____
   If no, note which lights do not work? _____
   If none of the brake lights operate what should be checked? _____
   _____

3. Locate the specification for the amount of brake pedal travel required to activate the brake light circuit.
   Spec _____

4. Measure the brake pedal travel necessary to activate the brake lights and note your results.
   _____

5. Is the brake light switch properly adjusted?   Yes _____   No _____

6. Explain the procedure to adjust the brake light switch. _____
   _____
   _____

INSTRUCTOR VERIFICATION:

Lab Worksheet 11-3

Name _____ Date _____ Instructor _____

Year _____ Make _____ Model _____

Brake Warning Lamp

1. Turn the ignition on and note the brake warning lamp.   On _____   Off _____

   If the lamp does not illuminate, what might this mean? _____
   _____

   What steps would you take to diagnose this problem? _____
   _____

2. Start the engine and note the warning lamp.   Remains on _____   Off _____

   List three possible causes for the lamp remaining on with the engine running.
   _____
   _____
   _____

3. If the lamp is off, carefully apply and release the parking brake to check warning light operation. The light should illuminate with the parking brake set and turn off when released.

   Light on _____   Light remains off _____

   If the light remains off, what would you need to check? _____
   _____

4. Is the vehicle is equipped with a brake fluid level sensor?   Yes _____   No _____

5. If the fluid sensor is built into the master cylinder cap, remove the cap and note if the brake warning lamp illuminates.   Yes _____   No _____

6. If the sensor is built into the reservoir, unplug the sensor and jump the two terminals together and note if the warning lamp illuminates.   Yes _____   No _____

7. If the light does not illuminate, what should be checked? _____
   _____

INSTRUCTOR VERIFICATION:

Lab Worksheet 11-4

Name _____ Date _____ Instructor _____

Year _____ Make _____ Model _____

Master Cylinder Inspection

1. Locate the master cylinder and note the following:
   Number of brake lines attached _____
   Location of reservoir _____
   Fluid level sensor   Yes _____   No _____

2. With the engine off, pump the brake pedal 10 times and hold.
   Does the pedal slowly sink to the floor?   Yes _____   No _____
   If yes, what can cause this? _____

3. While keeping pressure on the brake pedal, start the engine. Does the pedal sink to the floor?
   Yes _____   No _____
   If yes, what does this indicate? _____

4. Inspect the area where the master cylinder is attached to the brake booster.
   Is brake fluid or any wetness present?   Yes _____   No _____
   If yes, what does this indicate? _____

5. Inspect the brake line connections at the master cylinder.
   Is brake fluid or any wetness present?   Yes _____   No _____
   If yes, what does this indicate? _____

6. Based on your inspection, what is the condition of the master cylinder? _____
   _____
   _____

INSTRUCTOR VERIFICATION: _____

Lab Worksheet 11-5

Name _____ Date _____ Instructor _____

Year _____ Make _____ Model _____

Brake Bleeding

1. Before attempting to bleed the brake system, first locate and record the bleeding procedures outlined in the manufacturer's service information. _____
   _____
   _____

2. Locate the bleeder valves and make sure each will open. Do not force the bleeder valves as they break easily. Note any problems opening the bleeder valves. _____
   _____

3. If any or all bleeder valves are seized, what actions may be required to bleed the brake system?
   _____
   _____

4. Which type of bleeding procedure is to be used?
   Manual bleeding _____   Pressure bleeding _____
   Vacuum bleeding _____   Other _____
   Is a scan tool required?   Yes _____   No _____

5. Bleed the brake system following the manufacturer's procedures and note any problems you encounter.
   _____
   _____

   Instructor's Check _____

INSTRUCTOR VERIFICATION:

Drum Brake System Principles

Review Questions

1. Drum brakes use a set of brake _____ that expand outward against the inside of the rotating brake _____.

2. _____ pressure acts on the pistons in the wheel cylinders to press the shoes outward.

3. Explain the advantage drum brakes utilize to increase the amount of stopping power available compared to disc brakes. _____

4. List four disadvantages of drum brakes compared to disc brakes. _____

5. The _____ _____ holds the components of the drum brake assembly.

6. Brake _____ are curved pieces of metal onto which the friction lining is applied.

7. True or False: All four brake shoes are identical and can be installed on either axle and in either the front or rear position _____.

8. Compare and contrast a brake drum and a brake rotor. _____

9. The shoes are pulled back from the drum by the _____ springs.

10. Holddown springs and _____ hold the shoes to the backing plate.

11. Hydraulic pressure forces the two pistons inside of the _____ _____ outward and against the shoes.

12. Drum brakes use a _____ - _____ to maintain the shoe-to-drum clearance as the brakes wear.

13. The location of the _____ determines whether the brakes are servo or non-servo.

14. Brakes that use leverage from one shoe against the other shoe to increase brake application force are called:
    a. duo-servo brakes.
    b. self-energizing brakes.
    c. dual-servo brakes.
    d. all of the above.

15. Some drum brake designs use the self-adjuster as part of the _____ brake.

16. Non-servo brakes are also called _____ - _____ brakes.

17. The most common types of self-adjusters are _____ on one end and freely rotate on the other.

18. A _____ type self-adjuster is operated by using the parking brake _____.

19. The parking brake operation on a vehicle with rear drum brakes is partially dependent on how well the _____ are adjusted.

20. The parking brake is a _____ brake used primarily to lock the brakes when the vehicle is parked.

21. True or False: Only vehicles with manual transmissions really need to use the parking brake.

22. Parking brakes can be activated by:
    a. a hand-operated lever.
    b. an electrical input.
    c. a foot-operated pedal.
    d. all of the above.

23. Describe the two types of parking brake cables. _____
_____
_____
_____

24. Technician A says all parking brake systems are mechanically operated. Technician B says some vehicles use electrically operated parking brakes. Who is correct?

    a. Technician A
    b. Technician B
    c. Both A and B
    d. Neither A nor B

INSTRUCTOR VERIFICATION:

248 Chapter 12 Drum Brake System Principles

Activities

1. Identify the components of the drum brakes shown in Figure 12-1.

   a. _____ b. _____
   c. _____ d. _____
   e. _____ f. _____
   g. _____ h. _____
   i. _____ j. _____
   k. _____

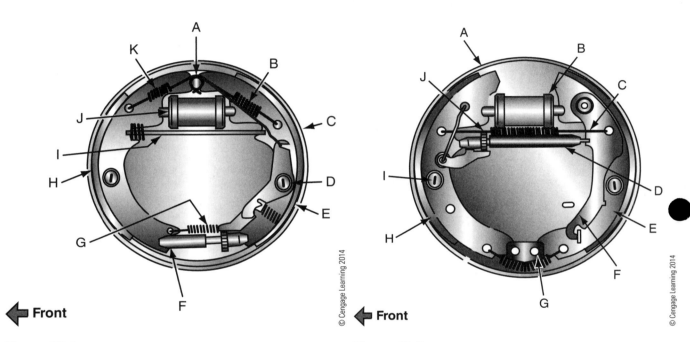

⬅ Front

Figure 12-1

⬅ Front

Figure 12-2

2. Identify the components of the drum brake shown in Figure 12-2.

   a. _____ b. _____
   c. _____ d. _____
   e. _____ f. _____
   g. _____ h. _____
   i. _____ j. _____

INSTRUCTOR VERIFICATION:

© 2014 Cengage Learning. All Rights Reserved. May not be scanned, copied or duplicated, or posted to a publicly accessible web site, in whole or in part.

3. Identify the components of the wheel cylinder shown in Figure 12-3.

   a. _____   b. _____
   c. _____   d. _____
   e. _____   f. _____

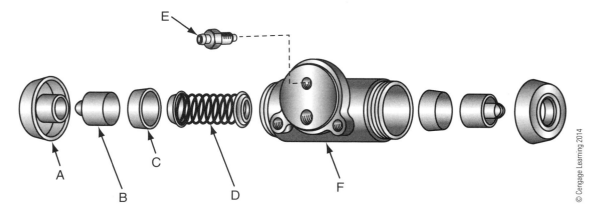

Figure 12-3

INSTRUCTOR VERIFICATION:

Lab Worksheet 12-1

Name _____ Date _____ Instructor _____

Year _____ Make _____ Model _____

Identify Brake Type and Components

Inspect the drum brakes and determine the following:

1. Drum brake type_____ Servo _____ Non-servo

2. Location of the anchor _____

3. Type of holddown springs   _____ Coil   _____ U-clip

    _____ Spring   _____ Other

4. Self-adjuster type   _____ Threaded   _____ Ratcheting

5. Describe how the self-adjuster is actuated. _____
   _____

6. How are the linings attached to the shoes?

    _____ Riveted   _____ Glued

INSTRUCTOR VERIFICATION:

CHAPTER 13
Drum Brake System Inspection and Service

Review Questions

1. Explain why it is important to use the proper brake tools and equipment. _____
   _____
   _____
   _____

2. A vacuum enclosure or a _____ _____ is used to clean brake dust from the brake assembly.

3. A vacuum or sink is used to trap airborne dust and _____ fibers that may be present in the linings.

4. A _____ spring tool is used to safely remove and install high-tension _____ springs.

5. To measure brake drum wear and out-of-round, a drum brake _____ is used.

6. Do not use _____ or _____ in place of the proper brake tools.

7. Before beginning to disassemble a drum brake assembly, locate a diagram or take a _____ of the components to help with reassembly.

8. List five service precautions you should follow before you begin to service the brakes.
   a. _____
   b. _____
   c. _____
   d. _____
   e. _____

253

Chapter 13 Drum Brake System Inspection and Service

9. Describe how to safely contain asbestos dust when working on brake systems.
   _____
   _____
   _____
   _____

10. Any time a vehicle is checked for a brake system complaint, _____ wheel brake assemblies should be inspected.

11. A full brake inspection, including the operation of the _____ brake and warning lights, should be performed any time a brake concern is present.

12. All of the following can cause drum brake noise except:
    a. excessive dust trapped in the drum.
    b. high metal content in the linings.
    c. shoe wear indicator.
    d. inadequate lubrication between the shoes and backing plate.

13. _____ is where the brake applies too quickly or with too much force, which causes the wheel to lock.

14. Technician A says an out-of-round drum may make the whole car shake when the brakes are applied. Technician B says a brake pulsation that disappears when only the parking brake is applied is caused by the front disc brakes. Who is correct?
    a. Technician A
    b. Technician B
    c. Both A and B
    d. Neither A nor B

15. A _____ brake pedal is usually caused by excessive shoe-to-drum clearance or _____ shoes and drums.

16. Before you attempt to remove the brake drum, check the _____ _____ for the procedure to remove the drum.

17. Technician A says all modern cars and trucks used floating drums. Technician B says floating drums require the rear axle to be removed to remove the drum. Who is correct?
    a. Technician A
    b. Technician B
    c. Both A and B
    d. Neither A nor B

18. Regarding Figure 13-1: Technician A says the retainer should be reinstalled after the brakes are inspected. Technician B says the retainer can be discarded once the drum is removed. Who is correct?

    a. Technician A
    b. Technician B
    c. Both A and B
    d. Neither A nor B

Figure 13-1

19. Describe how to remove a floating brake drum. _____
    _____
    _____

20. The three most common non-floating drum arrangements are the _____ bearing, _____ wheel bearing, and _____ -floating rear axle types.

21. A common method of removing full-floating drums involves unbolting the _____ flange from the hub and removing the axle.

22. Two technicians are discussing the brake dust shown in Figure 13-2: Technician A says the dust is wet from a leaking wheel cylinder. Technician B says the wet dust can be cleaned off and the shoes remain in service. Who is correct?

    a. Technician A
    b. Technician B
    c. Both A and B
    d. Neither A nor B

Figure 13-2

23. What is shown in Figure 13-3?
    a. Measuring brake shoe width
    b. Measuring lining thickness
    c. Adjusting the parking brake
    d. Checking shoe-to-drum clearance

Figure 13-3

24. Hard spots in the drum are caused by _____ the brakes, causing the metal to change under heat stress.

25. _____ in the drum facing, such as around lug holes, can occur from extreme stress or from a collision.

26. A _____ -mouthed drum is one in which the inside diameter is less than the diameter around the outside of the drum.

27. Drum brake _____ consists of the springs and related parts of the drum brake assembly.

28. What inspection is being performed in Figure 13-4?
    a. Wheel cylinder leak
    b. Axle seal leak
    c. Shoe lining thickness
    d. None of the above

Figure 13-4

29. Describe how to check parking brake operation. _____
    _____
    _____
    _____

30. Begin drum brake disassembly by cleaning as much brake _____ from the linings and hardware as possible with either a wet sink or brake dust vacuum.

31. If you are working on brakes that you are not familiar with, take a _____ of the brake before you begin disassembly.

32. Describe how and why should the backing plate be inspected? _____
    _____
    _____
    _____

33. Bonded linings should be replaced when the lining wears to a thickness of _____ inch or about _____ mm.

34. Many technicians recommend replacing the _____ and spring _____ when the shoes are replaced.

35. The brake drum should be replaced if it has which of the following defects?
    a. Cracks in the friction surface
    b. Exceeds maximum diameter
    c. Deep scoring
    d. All of the above

36. Before attempting to refinish a brake drum, first measure the drum diameter with a drum brake _____.

37. When refinishing a drum, what is the final step before reinstalling the drum on the vehicle?
    a. Applying a non-directional finish to the drum surface
    b. Washing the drum thoroughly inside to remove dust
    c. Lubricate the drum surface with high-temperature brake lubricant
    d. Reapplying drum balance weights removed during refinishing

38. List the steps in replacing a wheel cylinder. _____
   _____
   _____
   _____

39. Which of the following lubricants may be used when reassembling the drum brakes?
    a. Chassis grease
    b. WD-40
    c. Silicone spray
    d. None of the above

40. What action is shown in Figure 13-5?
    a. Measuring the brake drum diameter
    b. Determining shoe lining thickness
    c. Measuring to pre-adjust the shoes
    d. Measuring hub runout

Figure 13-5

41. A vehicle with a disc/drum brake system has a low service brake and the parking brake does not hold when applied. Technician A says the parking brake adjustment is incorrect. Technician B says the drum brake self-adjusters may not be working. Who is correct?

    a. Technician A
    b. Technician B
    c. Both A and B
    d. Neither A nor B

Activities

1. Match the drum brake concern with the most likely cause.

   Brake pulsation   Sticking parking brake cable

   Brake grab   Bent backing plate

   Brake noise   Fluid contamination

   Abnormal lining wear   Excessive dust buildup

   Brake drag   Out-of-round drum

INSTRUCTOR VERIFICATION:

Lab Worksheet 13-1

Name _____ Date _____ Instructor _____

General Drum Brake Inspection

BRAKE SYSTEM INSPECTION

Date: _____
Name: _____
Address: _____
City: _____ State: _____ Zip Code: _____
Year: _____ Make: _____ Model: _____
Mileage: _____ License: _____ VIN: _____ Eng: _____

Customer Interview:
_____
_____
_____
_____
_____

INSTRUCTOR VERIFICATION:

Chapter 13 Drum Brake System Inspection and Service

BRAKE INSPECTION CHECKLIST

Owner _____ Phone _____ Date _____

Address _____ VIN _____

Make _____ Model _____ Year _____ Lic. number _____ Mileage _____

Road test results		Yes	No	Right	Left	Underhood checks	OK	Not ok
Brake pulsation	Above 30 MPH					Brake fluid level		
	Below 30 MPH					Brake fluid moisture content		
	Bump steer					Brake fluid copper content		
						Master cylinder reservoir cover/gasket		
Steering wheel movement						Master cylinder		
Pull under braking						Lines and fittings		
Noise under braking						**Power booster**	OK	
ABS activation						Vacuum supply		
		OK	Not ok			Vacuum hose and check valve		
Power assist operation								
ABS light on/off		Yes	No					
Parking brake travel		OK	Not ok			Electric parking brake operation	OK	Not ok
Parking brake cables		OK	Not ok			comments		

Rotor thickness	Specs Frnt ____ Rear ____	**Rotor runout**	Specs Frnt ____ Rear ____
measured		measured	
RF ____ LF ____ RR ____ LR ____		RF ____ LF ____ RR ____ LR ____	
Rotor parallelisms	Specs Frnt ____ Rear ____	**Drum diameter**	Specs Frnt ____ Rear ____
measured		measured	
RF ____ LF ____ RR ____ LR ____		RF ____ LF ____ RR ____ LR ____

Brake hoses and lines	OK	Not	Drum condition	OK	Not
Caliper fluid leaks			Brake hoses and lines		
Hardware			Wheel cylinder fluid leaks		
Mounting bolts			Hardware		
Bracket bolts			Hold down springs		
Guides and pins			Return springs		
Boots			Self adjuster		
Clips, springs, other			Parking brake		
Lining thickness Spec Fnt ____ Rear ____			Clips		
measured			**Lining thickness** Spec Fnt ____ Rear ____		
RF ____ LF ____ RR ____ LR ____			Comments		

Other	OK	Not OK
wheel bearings		
control arm bushings		
strut rod bushings		
radius arm bushings		
Other		

INSTRUCTOR VERIFICATION:

Lab Worksheet 13-2

Name _____ Date _____ Instructor _____

Year _____ Make _____ Model _____

Drum Inspection

1. Inspect the outside of the brake drum for damage and note your findings. _____
   _____
   _____

2. With the drum still on the vehicle, spin the drum by hand, and note the following:

   Spins easily   Yes _____   No _____

   Continues to spin several times once spun Yes _____   No _____

   Gets tight and loose as it spins   Yes _____   No _____

   Noise when spinning   Yes _____   No _____

3. Remove the drum and inspect the friction surface for scoring, cracks, pitting, and bluing. Note your findings. _____
   _____

4. Locate the drum diameter specs and record.

   Max diameter _____

   Service limit or machine to spec _____

5. Using a drum micrometer, measure the drum and record your readings.

   Diameter _____

   Out-of-round _____

Chapter 13 Drum Brake System Inspection and Service

6. Examine the drum for signs of bell-mouthing, concave or convex wear, heat cracks, or checking. Note your findings. _____
   _____
   _____

7. Based on your inspection, what are your conclusions about the drum? _____
   _____
   _____

INSTRUCTOR VERIFICATION:

Lab Worksheet 13-3

Name _____ Date _____ Instructor _____

Year _____ Make _____ Model _____

Wheel Cylinder Inspection

1. Perform a visual inspection of the outside of the brake drum and the backing plate. Look for evidence of fluid leaking around the drum and backing plate. Note your findings.
   _____
   _____

2. Remove the brake drum. Describe the type of drum brakes installed on the vehicle.
   _____
   _____

3. Inspect the dust in the drum and on the shoes.

   Does the dust appear wet?   Yes _____   No _____

   If yes, what appears to be the most likely cause of the leak? _____

4. Carefully pull back the edge of one of the wheel cylinder's dust boots.

   Is fluid present in the boot?   Yes _____   No _____

   If yes, what does this indicate? _____

5. Carefully attempt to push a piston back into the cylinder bore.

   Does the piston move inward?   Yes _____   No _____

   If no, what does this indicate? _____

6. Inspect the brake line connection and the bleeder screws at the rear of the wheel cylinders. Describe their appearance. _____
   _____

7. Carefully attempt to loosen the bleeder screws. Do the bleeder screws open easily?

   Yes _____   No _____

   If no, what actions may be required to open the bleeder screws? _____
   _____

8. Based on your inspection what actions are required? _____

INSTRUCTOR VERIFICATION:

Lab Worksheet 13-4

Name _____ Date _____ Instructor _____

Year _____ Make _____ Model _____

Parking Brake Inspection

1. Describe the type and operation of the parking brake system used on this vehicle.
   _____
   _____
   _____

2. Locate and summarize the manufacturer's inspection procedures for the parking brake system.
   _____
   _____
   _____

3. Raise the vehicle and inspect the parking brake cables and components that are visible. Note your findings.
   _____
   _____

4. Have a helper slowly start to apply the parking brake while you watch the cables. Have the helper stop if the parking brake is very tight. Apply the brake and then release.

   Did the cables move?   Yes _____   No _____

   If no, what may this indicate? _____
   _____

   Did the parking brake release properly?   Yes _____   No _____

5. If the cables do not move or the brake did not release, locate the cable or part that is binding or is frozen. Note your findings. _____

6. If the cables move properly, apply the brake the specified amount according to the manufacturer's service procedures. Specified number of clicks: _____

   Does the parking brake set properly and hold?   Yes _____   No _____

7. Based on your inspection, what is the condition of the parking brake system? _____
_____
_____

INSTRUCTOR VERIFICATION:

… Chapter 13 Drum Brake System Inspection and Service   269

Lab Worksheet 13-5

Name _____ Date _____ Instructor _____

Year _____ Make _____ Model _____

Electric Parking Brake Inspection

1. Describe the type and operation of the electric parking brake system used on this vehicle.
   _____
   _____
   _____

2. Locate and summarize the manufacturer's inspection procedures for the parking brake system.
   _____
   _____
   _____

3. Have a helper apply the parking brake while you monitor the calipers. Apply the brake and then release.

   Did the brake apply?   Yes _____   No _____

   If no, what may this indicate? _____

   Did the parking brake release properly?   Yes _____   No _____

4. If the brake did not apply, note any warning or fault lights that are illuminated on the instrument panel. Note your findings. _____

5. If the brake did not apply, connect a scan tool to the vehicle's DLC. Navigate to the Brake System menu and check for stored diagnostic trouble codes. Are any current or history codes present?
   Yes _____   No _____

   If yes, record the stored codes. _____

6. Based on your inspection, what is the condition of the parking brake system? _____
   _____
   _____

INSTRUCTOR VERIFICATION: _____

CHAPTER 14

Disc Brake System Principles

Review Questions

1. Front disc brakes are _____ on all modern cars and light trucks, and disc brakes are often utilized for the _____ brakes as well.

2. Which of the following is not an advantage of disc brakes?
   a. Self-adjusting
   b. Increased fade resistance
   c. Decreased driver effort
   d. Self-cleaning of dust

3. Describe in your own words basic disc brake operation. _____
   _____
   _____
   _____

4. The _____ is the hydraulic output for the disc brake system.

5. Technician A says all disc brake systems use a caliper to push the brake pads out against the brake rotor. Technician B says some disc brake systems use multiple calipers to apply the brake pads against the rotor. Who is correct?
   a. Technician A
   b. Technician B
   c. Both A and B
   d. Neither A nor B

6. Which of the following is not a common type of disc brake caliper?
   a. Fixed
   b. Floating
   c. Fixed-sliding
   d. Sliding

271

Chapter 14 Disc Brake System Principles

7. Fixed calipers have at least _____ pistons.

8. In a fixed caliper design, how is the outer brake pad applied against the rotor?
    a. Leverage from the inner caliper piston
    b. Hydraulic pressure
    c. The caliper sliding backward
    d. None of the above

9. The piston seal, sometimes called a _____ seal, seals each piston in the caliper bore.

10. Caliper piston dust boots are _____ type seals, meaning they will expand and contract to cover the piston as the piston moves outward in the bore.

11. Describe how the piston is retracted back into the caliper bore when the brakes are released. _____
_____

12. All of the following statements about floating calipers is correct except:
    a. Floating calipers are smaller and less expensive than fixed calipers.
    b. Floating calipers use Newton's Third Law of Motion to operate.
    c. Floating calipers are often mounted on bushings or sleeves.
    d. Floating calipers use both inboard and outboard pistons.

13. Newton's Third Law of Motion states that for every _____ there is an _____ and _____ reaction.

14. For the _____ caliper to operate correctly, the caliper must be able to move freely on the mounting _____.

15. _____ calipers do not use mounting bolts or pins; instead they are mounted so that they can slide on the steering knuckle.

16. Which of the following is not a common caliper piston construction material?
    a. Cast iron
    b. Plastic
    c. Aluminum
    d. Steel

17. Floating calipers have mounting _____, pins, or bushing bores.

18. Explain the purpose of caliper hardware. _____
_____

19. Label the components of the brake pad shown in Figure 14-1 (a through e).

a. ____________________ b. ____________________

c. ____________________ d. ____________________

e. ____________________

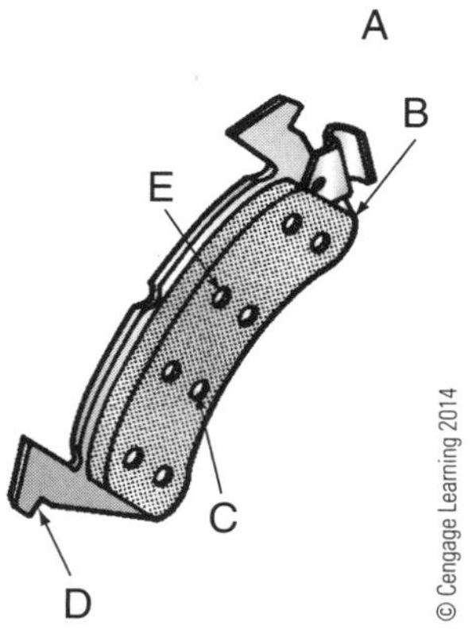

Figure 14-1

20. Many pads also use ____________________ to help reduce noise.

21. Rotors, also called ____, are mounted to the hub and rotate with the wheel and tire.

22. Label the parts of the brake rotor shown in Figure 14-2 (a through e).

a. ____________________ b. ____________________

c. ____________________ d. ____________________

e. ____________________

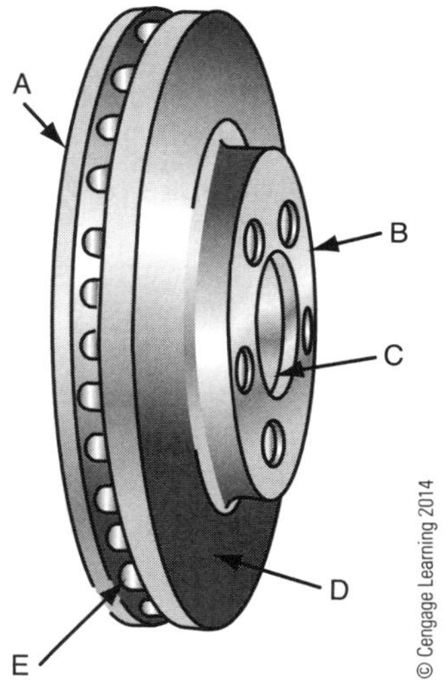

Figure 14-2

© 2014 Cengage Learning. All Rights Reserved. May not be scanned, copied or duplicated, or posted to a publicly accessible web site, in whole or in part.

Chapter 14 Disc Brake System Principles

23. The center of the rotor is also called the _____.

24. The overall rotor friction surface area is _____, but the contact area of the _____ is small.

25. Explain why nonvented brake rotors are not used on front disc brakes. _____

26. Technician A says all disc brake rotors are vented to improve heat dissipation. Technician B says rotor vents can be either straight or curved depending on the application. Who is correct?
    a. Technician A
    b. Technician B
    c. Both A and B
    d. Neither A nor B

27. Drilling or slotting rotors allows _____ to escape, which improves braking.

28. Brake _____ is the loss of braking performance, often as a result of heat.

29. Brake pad linings are a _____ between factors such as pad life, noise generation, and cold and hot coefficients of friction.

30. All of the following are brake pad friction material types except:
    a. Organic.
    b. Ceramic.
    c. Carbon fiber.
    d. Semimetallic.

31. Chamfering the leading edges of the brake pads helps decrease brake _____.

32. Describe two types of brake pad wear indicators and their operation. _____

33. Technician A says rear disc brake assemblies are similar in operation to front disc brake assemblies. Technician B says some rear disc brake systems include the parking brake system. Who is correct?
    a. Technician A
    b. Technician B
    c. Both A and B
    d. Neither A nor B

34. Describe the operation of an integral rear parking brake caliper. _____
_____
_____
_____

35. Rear disc brakes that have a set of small brake shoes enclosed by the brake rotor are called _____ - _____ - _____ systems.

36. True or False: Some vehicles have electric parking brake motors. _____

37. Discuss how regenerative braking operates. _____
_____
_____

Activities

1. Identify the types of the disc brake calipers shown in Figure 14-3, Figure 14-4, and Figure 14-5.

Figure 14-3 ____________________

Figure 14-4 ____________________

Figure 14-5 ____________________

© Cengage Learning 2014

Figure 14-3

© Cengage Learning 2014

Figure 14-4

© 2014 Cengage Learning. All Rights Reserved. May not be scanned, copied or duplicated, or posted to a publicly accessible web site, in whole or in part.

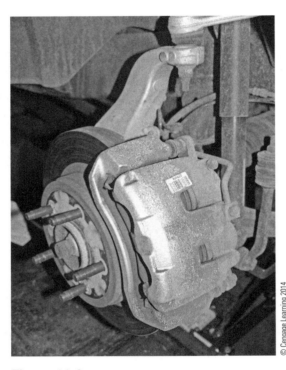

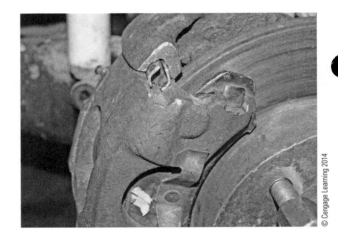

Figure 14-5

2. Identify the components of the disc brake caliper shown in Figure 14-6 (a through i).

a. _____ b. _____

c. _____ d. _____

e. _____ f. _____

g. _____ h. _____

i. _____

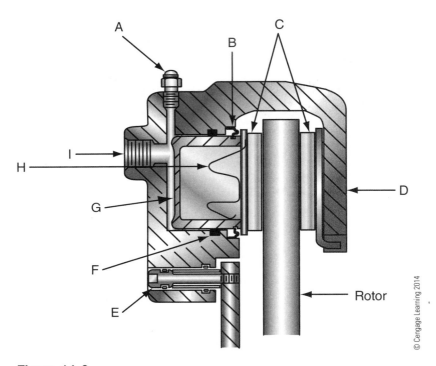

Figure 14-6

INSTRUCTOR VERIFICATION:

Lab Worksheet 14-1

Name _____ Date _____ Instructor _____

Year _____ Make _____ Model _____

Identify Disc Brake Types

1. Describe the type of front disc brake system used on this vehicle. _____
   _____

2. Caliper type _____ Fixed   _____ Floating   _____ Sliding

3. Number of caliper pistons   _____

   Caliper piston construction material _____

4. Rotor type   _____ Floating   _____ Trapped

5. Describe the type of rear disc brake system used on this vehicle. _____
   _____

6. Caliper type   _____ Fixed   _____ Floating   _____ Sliding

7. Number of caliper pistons   _____

   Caliper piston construction material _____

8. Rotor type   _____ Floating   _____ Trapped

9. Describe the type of parking brake system used. _____
   _____

INSTRUCTOR VERIFICATION: _____

CHAPTER 15

Disc Brake System Inspection and Service

Review Questions

1. Performing a brake job means more than just installing a set of brake _____.

2. Performing _____ system services, such as brake _____ replacement, is among the most common jobs technicians perform.

3. True or False: Brake work is not a common service performed by technicians. _____

4. Improper use of tools and taking shortcuts can lead not only to _____ damage, but can also cause personal _____.

5. It is important to always use the proper _____ for the job.

6. Many brake calipers are attached with either _____ or _____ head bolts.

7. What is the purpose of the tool shown in Figure 15-1?

Figure 15-1

8. To measure brake rotor wear, a rotor _____ is used.

9. A _____ is used to measure brake rotor runout.

10. List three precautions for working on the disc brake system. _____
_____
_____
_____

11. Wearing _____ helps protect your hands and also can help prevent getting new components dirty during installation.

12. Summarize three safety precautions for working on the brake system. _____
_____
_____
_____
_____
_____

13. List five methods of ensuring that brake work is performed safely and properly.
    a. _____
    b. _____
    c. _____
    d. _____
    e. _____

14. Technician A says replacing the brake pads can cause the brake pedal to feel different and can affect stopping performance. Technician B says flushing and filling the hydraulic system with a different brake fluid can affect brake pedal feel. Who is correct?
    a. Technician A
    b. Technician B
    c. Both A and B
    d. Neither A nor B

15. Because of the way in which disc brakes operate, they are prone to _____ issues.

16. Describe how noise is generated by the disc brake system. _____
_____
_____
_____

Chapter 15 Disc Brake System Inspection and Service

17. What is the purpose of the component shown in Figure 15-2?

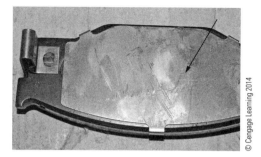

Figure 15-2

18. A rhythmic sound like a wire brush against metal can be caused by rotor _____ or _____ variation.

19. List five possible causes of disc brake noise.
    a. _____
    b. _____
    c. _____
    d. _____
    e. _____

20. The most common causes of disc brake _____ are restricted hydraulic hoses and _____ caliper pistons.

21. Worn and loose control arm _____ can cause the vehicle to pull when braking.

22. Describe how to check for restricted hoses and seized calipers. _____

23. All of the following can cause disc brake pulsation concerns except:
    a. Out-of-round rotor
    b. Excessive thickness variation
    c. Hub runout
    d. Loose wheel bearing

24. Brake _____ refers to a brake that applies with too much braking force, causing the wheel to lock up easily.

25. Brake _____ occurs when a brake remains applied after the brake pedal is released.

26. Which of the following is not a common disc brake complaint?
    a. Noise
    b. Pulsation
    c. Pulling
    d. Firm pedal feel

27. A vehicle has uneven pad wear on the left and right side brake pads. Technician A says a caliper may be sticking and not releasing properly. Technician B says excessive rust between the pads and the caliper bracket may be the cause. Who is correct?
    a. Technician A
    b. Technician B
    c. Both A and B
    d. Neither A nor B

28. Begin your brake inspection with a _____ with the driver of the vehicle.

29. Both the red _____ light and the _____ light should illuminate during engine start and bulb check and then go out after a few seconds.

30. A leaking caliper must be either rebuilt or _____.

31. Inspect the hoses for signs of _____ or contact with other components.

32. Brake pads should be replaced when the lining reaches approximately _____ (inch or mm) in thickness.

33. Technician A says a pad lining will wear faster once it is half worn out. Technician B says a pad will wear at the same rate regardless of how much wear is on the linings. Who is correct?
    a. Technician A
    b. Technician B
    c. Both A and B
    d. Neither A nor B

34. Uneven pad wear can indicate that the pads are not _____ properly and are not fully releasing from the applied position against the rotor.

35. Before starting to _____ the brakes, first ensure you understand how the brake is assembled.

36. What action is being performed in Figure 15-3?

Chapter 15 Disc Brake System Inspection and Service

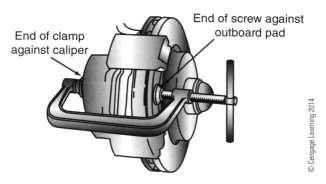

Figure 15-3

37. Do not allow the _____ to hang on the brake as this can damage the hose.

38. When you are ready to reassemble the brakes, use a small amount of brake _____ on places where the pads or calipers move.

39. Before you start and move the vehicle after the pads have been replaced, _____ the brake pedal several times until the pedal pumps up and is _____.

40. Never let the _____ be the first person to drive the vehicle after the brakes have been serviced.

41. Describe pad break-in and why it is important for the pads and rotors. _____
_____
_____
_____

42. Brake rotors should be free of:
    a. Deep scoring.
    b. Glazing.
    c. Cracks.
    d. All of the above.

43. Technician A says a rotor may rust from the inside out, causing the rotor to separate through the vents. Technician B says the inner surface of the rotor should be closely inspected for rust. Who is correct?
    a. Technician A
    b. Technician B
    c. Both A and B
    d. Neither A nor B

44. Which of the following is not a brake rotor wear specification?
    a. Maximum thickness
    b. Machine-to or minimum refinish thickness
    c. Runout
    d. Parallelism

45. The machine-to spec is often _____ (inches or mm) to _____ (inches or mm) larger than the discard specification.

46. What measurement is shown in Figure 15-4 and why is it important?
    _____
    _____
    _____
    _____

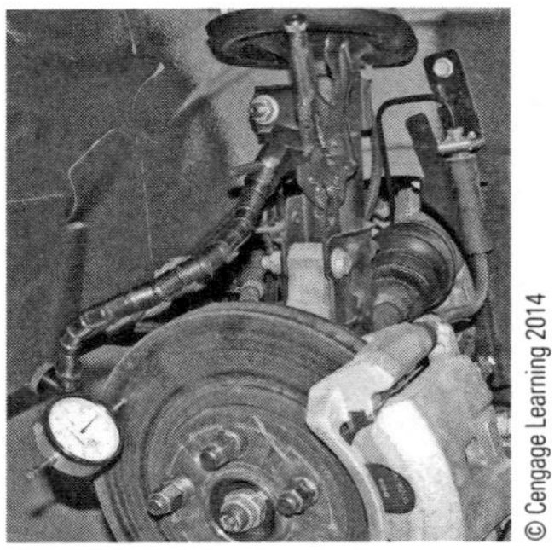

Figure 15-4

47. Explain the differences between off-car and on-car rotor machining. _____
    _____
    _____

48. Why do some vehicles use on-car brake rotor machining? _____
    _____
    _____
    _____

49. Explain how the caliper piston is retracted on an integral parking brake caliper.
    _____
    _____

50. Explain the final checks you should perform once the brake pads have been replaced.
   _____
   _____
   _____

Activities

1. Place the following brake pad service steps in proper order.

   a. Measure rotor thickness

   b. Loosen and remove caliper pins

   c. Set caliper piston

   d. Remove pads

   e. Remove caliper and hang from wire

   f. Reinstall and torque

   g. Lube pad guides

   h. Lube caliper pins

2. Match the following complaints with their most likely causes.

   Pulsation   Restricted brake hose

   Squealing   Thickness variation

   Drag   Stuck caliper piston

   Abnormal pad wear   Binding caliper pins

   Pull   Hydraulic leak

INSTRUCTOR VERIFICATION:

Lab Worksheet 15-1

Name _____ Date _____ Instructor _____

Disc Brake Inspection

BRAKE SYSTEM INSPECTION

Date: _____
Name: _____
Address: _____
City: _____ State: _____ Zip Code: _____
Year: _____ Make: _____ Model: _____
Mileage: _____ License: _____ VIN: _____ Eng: _____

Customer Interview:
_____
_____
_____
_____
_____
_____

INSTRUCTOR VERIFICATION:

Chapter 15 Disc Brake System Inspection and Service

BRAKE INSPECTION CHECKLIST

Owner _____ Phone _____ Date _____

Address _____ VIN _____

Make _____ Model _____ Year _____ Lic. number _____ Mileage _____

Road test results		Yes	No	Right	Left	Underhood checks	OK	Not ok
Brake pulsation	Above 30 MPH					Brake fluid level		
	Below 30 MPH					Brake fluid moisture content		
	Bump steer					Brake fluid copper content		
Steering wheel movement						Master cylinder reservoir cover/gasket		
Pull under braking						Master cylinder		
Noise under braking						Lines and fittings		
ABS activation						**Power booster**	**OK**	
Power assist operation		OK	Not ok			Vacuum supply		
ABS light on/off		Yes	No			Vacuum hose and check valve		
Parking brake travel		OK	Not ok			Electric parking brake operation	OK	Not ok
Parking brake cables		OK	Not ok			comments		

Rotor thickness	Specs Frnt _____ Rear _____	Rotor runout	Specs Frnt _____ Rear _____
measured		measured	
RF ____ LF ____ RR ____ LR ____		RF ____ LF ____ RR ____ LR ____	

Rotor parallelisms	Specs Frnt _____ Rear _____	Drum diameter	Specs Frnt _____ Rear _____
measured		measured	
RF ____ LF ____ RR ____ LR ____		RF ____ LF ____ RR ____ LR ____	

Brake hoses and lines	OK	Not	Drum condition	OK	Not
Caliper fluid leaks			Brake hoses and lines		
Hardware			Wheel cylinder fluid leaks		
Mounting bolts			Hardware		
Bracket bolts			Hold down springs		
Guides and pins			Return springs		
Boots			Self adjuster		
Clips, springs, other			Parking brake		
Lining thickness Spec Fnt _____ Rear _____			Clips		
measured			**Lining thickness** Spec Fnt _____ Rear _____		
RF ____ LF ____ RR ____ LR ____			Comments		

Other	OK	Not OK
wheel bearings		
control arm bushings		
strut rod bushings		
radius arm bushings		
Other		

INSTRUCTOR VERIFICATION:

Lab Worksheet 15-2

Name _____ Date _____ Instructor _____

Year _____ Make _____ Model _____

Identify Pad Wear

1. Describe the type of disc brake system being inspected. _____
   _____

2. Check for signs of abnormal pad and rotor wear by performing a visual inspection. Note your findings.
   _____
   _____

3. Remove the brake caliper from the knuckle or bracket to gain access to the pads. Describe what steps you used to remove the caliper. _____
   _____

4. Carefully remove the pads. Inspect the lining areas and note any of the following conditions:

   a. Excessive wear   Yes _____   No _____

   Note lining thickness _____   Note minimum lining spec _____

   b. Cracks   Yes _____   No _____

   c. Glazing   Yes _____   No _____

   d. Uneven wear   Yes _____   No _____

   If yes, describe the wear _____

   e. Other _____

5. Based on your inspection, what actions are necessary? _____
   _____
   _____

INSTRUCTOR VERIFICATION:

Lab Worksheet 15-3

Name _____ Date _____ Instructor _____

Year _____ Make _____ Model _____

Measure Brake Rotor Thickness and Parallelism

1. Locate and record the rotor specs.

    Nominal thickness _____   Minimum thickness _____

    Machine-to _____   Parallelism _____

2. Place the rotor micrometer in the middle of the rotor surface and note the reading.

    Micrometer reading _____

3. How does the rotor measurement compare to the minimum thickness spec?

    Over spec _____   Under spec _____   Within spec _____

4. Next, measure the rotor thickness at eight places around the friction surface and record your readings.

    _____   _____   _____   _____

    _____   _____   _____   _____

    Subtract the smallest number from the largest to find the parallelism difference.

    Parallelism measured _____

5. How does this measurement compare to the spec?

    Over spec _____   Under spec _____   Within spec _____

6. What causes excessive rotor parallelism? _____
_____

7. Based on your inspection, what is the condition of this rotor, and should it be placed back into service?

   _____

   _____

   _____

INSTRUCTOR VERIFICATION:

Lab Worksheet 15-4

Name _____ Date _____ Instructor _____

Year _____ Make _____ Model _____

Measure Rotor Runout

1. Locate and record the rotor runout specs.   Total runout _____

2. Install a dial indicator to measure rotor runout. Turn the rotor slowly and note the total needle movement.
   _____

3. How does this measurement compare to specs?

   Over spec _____   Under spec _____   Within spec _____

4. What are three possible causes for excessive rotor runout? _____
   _____
   _____

5. If the total indicated runout is greater than spec, note the location of the rotor to the hub, remove the rotor and reinstall it 180 degrees from its original position, and remeasure runout. Runout measured
   _____

6. If the runout decreased, what does this indicate? _____
   _____

7. If the runout remained excessive, remove the rotor and measure the runout of the hub. Runout measured
   _____

8. What is indicated by this measurement? _____
   _____

9. Based on your inspection, what is the reason for the runout and what should be done to correct the problem? _____
   _____
   _____
   _____

INSTRUCTOR VERIFICATION:

CHAPTER 16

Antilock Brakes, Electronic Stability Control, and Power Assist

Review Questions

1. Antilock brake systems use a _____ and _____ to monitor wheel speeds.

2. Vehicle stability control systems are designed to work with the antilock brake system to reduce _____ and _____.

3. Technician A says ABS is designed to decrease stopping distances in all braking situations. Technician B says ABS is designed to allow the driver to maintain control by preventing wheel lock. Who is correct?
   a. Technician A
   b. Technician B
   c. Both A and B
   d. Neither A nor B

4. ABS prevents wheel lock by which of the following actions?
   a. Applying reduced pressure to the wheel brake cylinders
   b. Applying increased pressure to the wheel brake cylinders
   c. Applying, holding, and releasing the pressure to the wheel brake cylinders
   d. None of the above

5. ABS and traction control systems cannot overcome which of the following factors?
   a. Insufficient traction between the tire and the ground
   b. Inoperative or faulty brake system components
   c. Gross driver negligence
   d. All of the above

6. The amount of traction between the tire and the ground is called tire _____.

7. _____ _____ _____ monitor the rotational speed of each wheel.

8. Which of the following components receives input and controls ABS operation?
   a. Wheel speed sensors
   b. EBCM
   c. Electro-hydraulic control unit
   d. Powertrain control module

9. EBCM stands for: _____

10. List the three modes in which the EBCM may command the hydraulic control unit.
    _____

11. The _____ usually contains electric motors, solenoids, and valves that control the flow of brake fluid.

12. Technician A says wheel speed sensors can produce either analog or digital signals. Technician B says all wheel speed sensors produce analog AC voltage signals. Who is correct?
    a. Technician A
    b. Technician B
    c. Both A and B
    d. Neither A nor B

13. Vehicles with ABS but without traction or stability control are being discussed: Technician A says the ABS only monitors wheel speed when the brakes are applied. Technician B says the ABS monitors wheel speed whenever the vehicle is moving. Who is correct?
    a. Technician A
    b. Technician B
    c. Both A and B
    d. Neither A nor B

14. Isolate mode is also called:
    a. Release mode.
    b. Hold mode.
    c. Reapply mode.
    d. Pressure increase mode.

15. During release mode, the EBCM commands pressure release so the wheel can _____.

16. Explain in detail the sequence of steps the ABS uses to control wheel lockup.
    _____
    _____
    _____
    _____

17. Never open a _____ line or _____ screw with the ignition on, as this can cause injury from the release of high-pressure fluid.

18. Integral ABS units use high-pressure _____, which must be discharged before any service is performed.

19. The first step in diagnosing an ABS concern is to _____ and _____ the customer complaint.

20. Explain why an inspection of the brake system components is important when diagnosing an ABS problem. _____
_____
_____

21. To flush and bleed the ABS hydraulic system, a _____ _____ may be required.

22. Describe the purpose of the traction control system. _____
_____
_____

23. List five requirements of vehicle stability control systems.
    a. _____
    b. _____
    c. _____
    d. _____
    e. _____

24. Describe the process of diagnosing a fault with the ABS and/or VSC system. _____
_____
_____
_____
_____

25. There are two types of external power brake assist systems, _____ assist and _____ assist.

26. Vacuum is air pressure that is less than _____ pressure.

27. Explain the purpose of the vacuum check valve. _____
_____

28. Some engines use a _____ pump to supply vacuum to the booster.

29. Explain how pressure differential is used to provide power assist in a vacuum booster.

   _____
   _____
   _____
   _____

30. Describe how atmospheric pressure, vacuum, and booster diaphragm size contribute to the amount of power assist generated by the vacuum booster. _____

   _____
   _____
   _____
   _____

31. Hydraulic power brake boosters may use fluid from the _____ _____ pump or an _____ pump.

32. Technician A says that a loose or missing power steering belt can affect the operation of a Hydro-boost system. Technician B says a faulty power steering pump can cause excessive brake pedal effort on a Hydro-boost system. Who is correct?

   a. Technician A
   b. Technician B
   c. Both A and B
   d. Neither A nor B

33. A vehicle with a vacuum power assist system has a very hard brake pedal and requires excessive effort to stop. All of the following can be the problem except:

   a. Leaking power booster diaphragm
   b. Misadjusted booster pushrod
   c. Blocked vacuum supply hose
   d. Leaking vacuum supply hose

34. When testing a vacuum power assist unit: Technician A says the brake pedal should slowly sink to the floor under light pressure when the engine is started. Technician B says the brake pedal should drop a couple of inches under light pressure once the engine is started. Who is correct?

   a. Technician A
   b. Technician B
   c. Both A and B
   d. Neither A nor B

35. Technician A says servicing brake systems on some hybrid vehicles requires following special service precautions and procedures. Technician B says some hybrid vehicles may apply the brakes even if the engine and ignition system are turned off. Who is correct?

    a. Technician A
    b. Technician B
    c. Both A and B
    d. Neither A nor B

Activities

1. Describe the purpose of the component shown in Figure 16-1.

_____

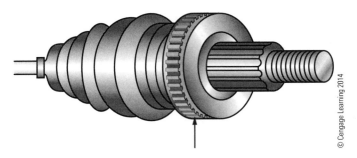

Figure 16-1

2. Label the components of the ABS system shown in Figure 16-2 (a through g).

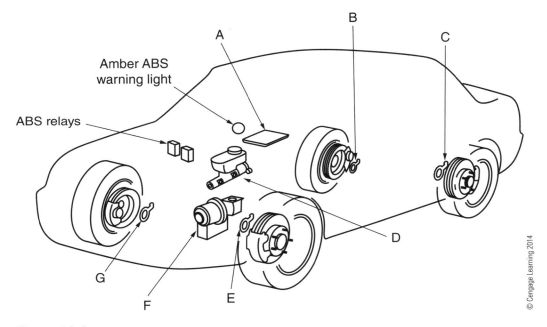

Figure 16-2

3. Label the components of the integral ABS system shown in Figure 16-3 (a through g).

Chapter 16 Antilock Brakes, Electronic Stability Control, and Power Assist 303

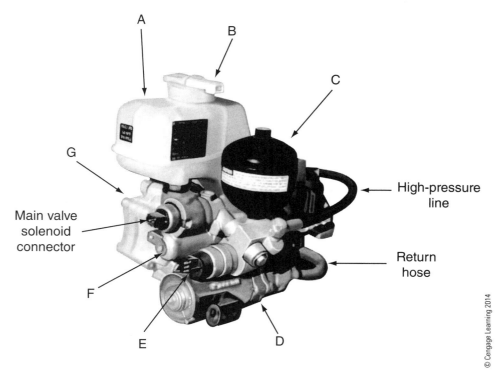

Figure 16-3

4. Label the parts of the vacuum booster shown in Figure 16-4 (a through g).

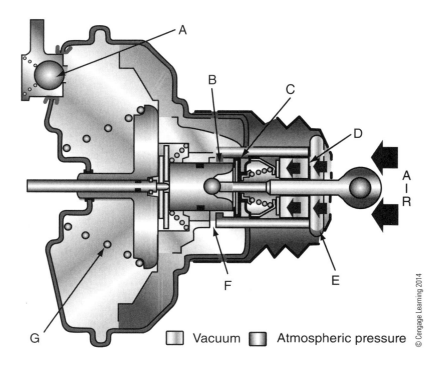

Figure 16-4

Lab Worksheet 16-1

Name _____ Date _____ Instructor _____

Year _____ Make _____ Model _____

ABS Inspection

1. Turn the ignition on and note the ABS light; does the light illuminate?

   Yes _____   No _____

   If no, what can this indicate? _____

2. Turn the engine on and note the ABS light; does the light stay on after several seconds?

   Yes _____   No _____

   If yes, what does this indicate? _____

3. If the ABS light remains on, refer to the manufacturer's service information to determine how to retrieve stored diagnostic trouble codes. Summarize the procedure. _____
   _____
   _____
   _____
   _____

4. Retrieve any codes and record them here. _____

5. What is indicated by the code(s)? _____
   _____
   _____

INSTRUCTOR VERIFICATION:

Lab Worksheet 16-2

Name _____ Date _____ Instructor _____

Year _____ Make _____ Model _____

Vehicle Modifications

1. Turn the ignition on and note the ABS light; does the light illuminate?

    Yes _____   No _____

    If no, what can this indicate? _____

2. Turn the engine on and note the ABS light; does the light stay on after several seconds?

    Yes _____   No _____

    If yes, what does this indicate? _____

3. Locate the vehicle tire placard and note the correct wheel and tire sizes.
   _____

4. Inspect the vehicle and note the tire sizes installed.

    RF tire _____   LF tire _____

    RR tire _____   LR tire _____

5. Are the correct tires installed on the vehicle?

    Yes _____   No _____

    If no, determine the height of the factory-installed wheel and tire.

    Factory wheel/tire height _____   Circumference _____

6. Measure the height of the wheel and tire installed on the vehicle.

    Installed wheel/tire height _____   Circumference _____

7. How can having nonfactory wheels and tires affect the operation of the ABS? _____

INSTRUCTOR VERIFICATION:

Lab Worksheet 16-3

Name _____ Date _____ Instructor _____

Year _____ Make _____ Model _____

WSS Testing

1. Turn the ignition on and note the ABS light; does the light illuminate?

   Yes _____   No _____

   If no, what can this indicate? _____

2. Turn the engine on and note the ABS light; does the light stay on after several seconds?

   Yes _____   No _____

   If yes, what does this indicate? _____

3. Determine the type of wheel speed sensor used on this vehicle.

   Passive _____   Active _____

   Explain the differences between passive and active sensors. _____
   _____
   _____

4. Describe three ways to test a passive WSS.
   a. _____
   b. _____
   c. _____

5. Describe two ways to test an active WSS.
   a. _____
   b. _____

6. Test a passive WSS provided by your instructor. Record your results. _____
   _____

Chapter 16 Antilock Brakes, Electronic Stability Control, and Power Assist

7. Based on your testing, what can you determine about this sensor? _____
   _____
   _____

8. Connect a scan tool and navigate to the ABS data. Display WSS speeds and drive the vehicle in a straight line at low speed. Do all sensors display the same speeds?

   Yes _____   No _____

   If no, what may this indicate? _____

9. Based on your inspection, what is the condition of the wheel speed sensors on this vehicle? _____
   _____
   _____

INSTRUCTOR VERIFICATION:

Lab Worksheet 16-4

Name _____ Date _____ Instructor _____

Identify Brake Assist Types

1. Year _____ Make _____ Model _____

   Determine the type of brake assist used in this vehicle.

   Vacuum boost _____   Hydraulic _____   ABS _____

2. Year _____ Make _____ Model _____

   Determine the type of brake assist used in this vehicle.

   Vacuum boost _____   Hydraulic _____   ABS _____

3. Year _____ Make _____ Model _____

   Determine the type of brake assist used in this vehicle.

   Vacuum boost _____   Hydraulic _____   ABS _____

4. Why do you think that the vehicles you inspected use that type of power brake assist?

   _____
   _____
   _____

INSTRUCTOR VERIFICATION:

Chapter 16 Antilock Brakes, Electronic Stability Control, and Power Assist 313

Lab Worksheet 16-5

Name _____ Date _____ Instructor _____

Year _____ Make _____ Model _____

Vacuum Booster Operation

1. Perform a visual inspection of the vacuum booster and vacuum hose, note your findings. _____
   _____
   _____

2. With the engine off, pump the brake pedal until the vacuum reserve is depleted. With your foot holding the brake pedal down, start the engine and note how the brake pedal responds. _____
   _____

3. Next, pump the brake pedal several times and note how it feels.

   Very hard to depress   Yes _____   No _____

   Moves down easily   Yes _____   No _____

4. With the brake depressed, listen for a hissing noise inside the vehicle.

   Is there a hiss present with the brake applied?   Yes _____   No _____

   What may a hiss with the brake applied indicate? _____

5. Turn the engine off and press the brake pedal down once. Does the pedal move down with normal pressure applied?   Yes _____   No _____

   If the pedal is very hard to press, what might this indicate? _____

6. Based on your inspection, what is the condition of the brake booster? _____
   _____
   _____

INSTRUCTOR VERIFICATION:

© 2014 Cengage Learning. All Rights Reserved. May not be scanned, copied or duplicated, or posted to a publicly accessible web site, in whole or in part.

CHAPTER 17

Electrical/Electronic System Principles

Review Questions

1. Current trends have been to either enhance or replace ___mechanical___ components with electrical components.

2. The term ___current electrical___ applies to electrical circuits that contain bulbs, motors, and other devices.

3. The term ___electronic___ refers to integrated circuits and computers that are not directly tested.

4. Give a brief explanation of electricity. ___The flow of electrons (amperage)___

5. Label the parts of the atom shown in Figure 17-1 (a through d).
   a. ___Orbital rings/shell___ b. ___Proton___
   c. ___Neutron___ d. ___Electron___

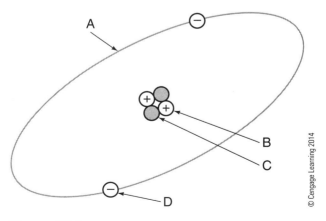

Figure 17-1

6. A copper atom, in its normal state with 29 protons and electrons, is called electrically balanced.

7. Static electricity is the term used to describe the accumulation of an electrical charge by a substance not normally conductive to electricity.

8. Define voltage. The amount of electrical potential, electrical energy something has

9. The amount of voltage being used by a component, wire, or connection is called Voltage drop.

10. Define amperage. Flow or Rate of flow of electricity

11. Define ohms. The electrical resistance in the flow of electricity

12. The relationship commonly used to describe electrical behavior is called Ohms law.

13. Place each letter in the correct location in Figure 17-2.

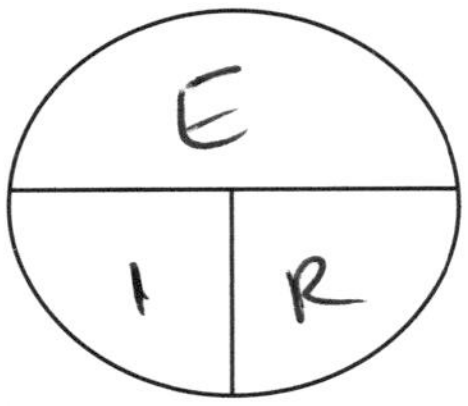

I
R
E

© Cengage Learning 2014

Figure 17-2

14. Explain what functions can and cannot be performed by volts, amps, and ohms using Ohm's law. As long as 2 of the other values are defined,

15. Define watts. The consumption and production of electrical power.

16. Describe alternating current. Has both positive and negative flow of current.

17. All of the following statements about AC voltage are correct except:
   a. AC has positive voltages.
   b. AC has zero voltage.
   c. AC changes between positive and negative voltage.
   d. AC eliminates negative voltages.

© 2014 Cengage Learning. All Rights Reserved. May not be scanned, copied or duplicated, or posted to a publicly accessible web site, in whole or in part.

18. All of the following statements about DC voltage are correct except:
    a. DC alternates between positive and negative. *(circled)*
    b. DC is used by battery powered devices.
    c. DC voltages can vary.
    d. Some devices require AC to be turned into DC to operate.

19. Explain several advantages and disadvantages of using AC and DC. AC is easy to generate. Electronic components only run on DC, so AC has to be converted to DC. DC can be stored.

20. *Circuit* is a term that refers to a complete electrical __push__ from power to ground.

21. A __circuit__ is a path for electrons to flow from a source of higher electrical potential to a source of lower potential.

22. List five components of a complete circuit.
    a. Source
    b. Conductors
    c. Load
    d. Circuit control
    e. Circuit protection

23. Label the parts of the circuit shown in Figure 17-3 (a through f).
    a. Fuse
    b. Switch
    c. Conductor
    d. Load
    e. Ground conductor
    f. Source of power / battery

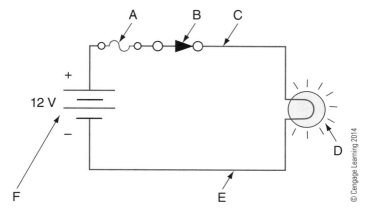

Figure 17-3

Chapter 17 Electrical/Electronic System Principles

24. The battery is the source of high electrical __potential__ for the circuit.

25. Negative or ground refers to the __return__ path of the circuit.

26. Anything that consumes electrical power is a __load__.

27. __Conductors__ are the wires, terminals, and connectors that connect the parts of the circuit together.

28. Switches are types of circuit __controls__, used to turn the circuit on and off.

29. List three forms of circuit protection devices.
    a. __Fuse__
    b. __Circuit breakers__
    c. __Fusable links__

30. List the components of a basic circuit. _____

31. Define the principles of a series circuit. _____

32. Define the principles of a parallel circuit. _____

33. Automotive circuits are arranged as _____-_____ circuits.

34. Fuses and circuit breakers are types of:
    a. Control devices.
    b. Loads.
    c. Protection devices.
    d. Conductors.

35. Circuit protection devices are used to limit the amount of _____ in a circuit and prevent wire and component damage.

36. A vehicle has a blown 10 A fuse. Technician A says to replace the fuse with a higher amperage rated fuse to prevent the fuse from blowing again. Technician B says if the fuse continues to blow, a fault exists with the fuse. Who is correct?
    a. Technician A
    b. Technician B
    c. Both A and B
    d. Neither A nor B

37. Circuit breakers are often used in circuits with _____ because of sometimes sudden surges in current flow.

38. Multiple fuses are located together in a fuse box or _____ box.

39. A digital multimeter can be used to measure:
    a. Amps.
    b. Ohms.
    c. Volts.
    d. All of the above.

40. A test light is most commonly used to test:
    a. Amps.
    b. Ohms.
    c. Volts.
    d. All of the above.

41. Technician A says unwanted resistance in a circuit can cause a component, such as a light bulb, not to operate. Technician B says unwanted resistance in a circuit will cause the fuse to blow. Who is correct?
    a. Technician A
    b. Technician B
    c. Both A and B
    d. Neither A nor B

42. List four possible causes for a bulb not to light up.
    a. _____
    b. _____
    c. _____
    d. _____

43. A reading of 0 volts means that there is no difference in _____ between the two meter leads.

44. DMMs only measure what is present _____ the ends of the positive and negative leads.

45. Electrical current flow generates a _____ field around a wire.

46. Explain how electromagnetism is used in modern vehicles. _____
_____
_____
_____

47. A relay is use to:
   a. Control circuit current flow levels.
   b. Control high current flow components.
   c. Protect a circuit.
   d. All of the above.

48. Explain the operation of a relay. _____
_____
_____
_____

49. Explain the basic operation of a DC motor. _____
_____
_____
_____
_____

50. Describe how the AC generator produces electricity to power the electrical system.
_____
_____
_____
_____

51. When referring to wire gauge, the larger the number of the gauge:
   a. The larger the wire diameter.
   b. The smaller the wire diameter.
   c. The more expensive the wire construction materials.
   d. None of the above.

52. A wiring _____ is a bundle of wires and connectors.

Chapter 17 Electrical/Electronic System Principles 321

53. Technician A says terminals are the parts of a harness that connect two wires together. Technician B says terminals contain the connectors that connect two wires together. Who is correct?
    a. Technician A
    b. Technician B
    c. Both A and B
    d. Neither A nor B

54. Explain what is meant by a common ground. _____
_____
_____

55. The basis of all electronics is the _____.

56. All of the following are types of electronic devices except:
    a. Diodes.
    b. LEDs.
    c. Transformers.
    d. Transistors.

57. Explain what a diode is and its common uses in the automobile. _____
_____
_____
_____

58. _____ are solid-state switches.

59. Explain what ESD is and how to prevent against accidental ESD damage.
_____
_____
_____
_____

Activities

1. Solve the series circuits in Figure 17-4 through Figure 17-8.

   a. Explain the relationship between current flow and resistance in the circuit. _____
   _____
   _____
   _____

   b. Describe the properties of series circuits. _____
   _____
   _____
   _____

   c. Why do you think series circuits are not generally used in automotive applications?
   _____
   _____
   _____
   _____

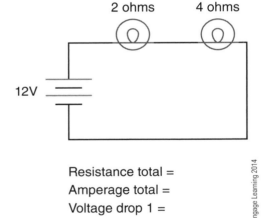

Resistance total =
Amperage total =
Voltage drop 1 =
Voltage drop 2 =

Figure 17-4

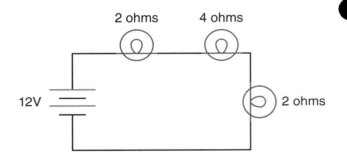

Resistance total =
Amperage total =
Voltage drop 1 =
Voltage drop 2 =
Voltage drop 3 =

Figure 17-5

INSTRUCTOR VERIFICATION:

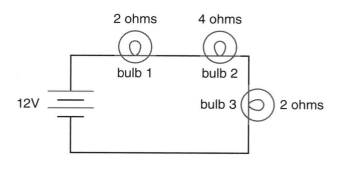

Resistance total =
Amperage total =
Voltage drop 1 =
Voltage drop 2 =
Voltage drop 3 =

Wattage of bulb 1 =
Wattage of bulb 2 =
Wattage of bulb 3 =
Total circuit wattage =

Figure 17-6

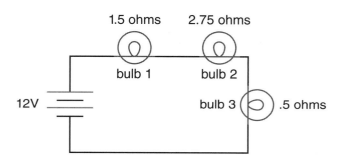

Resistance total =
Amperage total =
Voltage drop 1 =
Voltage drop 2 =
Voltage drop 3 =

Wattage of bulb 1 =
Wattage of bulb 2 =
Wattage of bulb 3 =
Total circuit wattage =

Figure 17-7

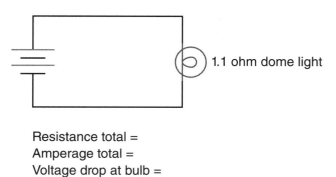
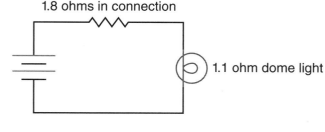

Resistance total =
Amperage total =
Voltage drop at bulb =

Resistance total =
Amperage total =
Voltage drop at resistance =

Figure 17-8

INSTRUCTOR VERIFICATION:

2. Solve the parallel circuits in Figure 17-9 through Figure 17-11.

   a. Describe the properties of parallel circuits. _____
   _____
   _____

   b. Explain how parallel circuits differ in operation from series circuits. _____
   _____
   _____
   _____

   c. What happens to current flow as branches are added to a parallel circuit? _____
   _____
   _____
   _____

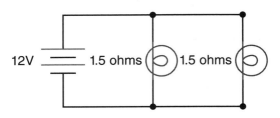

Resistance total =
Amperage through bulb 1 =
Amperage through bulb 2 =
Total amperage =
Total voltage drop =

Figure 17-9

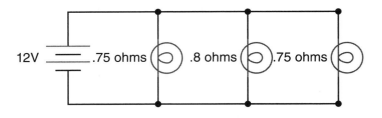

Resistance total =
Amperage through bulb 1 =
Amperage through bulb 2 =
Amperage through bulb 3 =

Total amperage =
Total voltage drop =

Figure 17-10

INSTRUCTOR VERIFICATION:

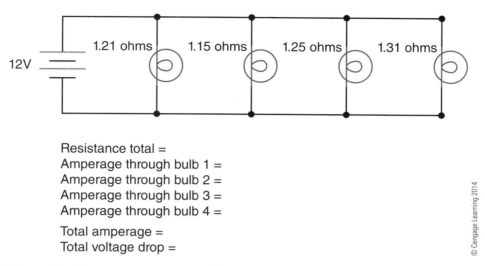

Resistance total =
Amperage through bulb 1 =
Amperage through bulb 2 =
Amperage through bulb 3 =
Amperage through bulb 4 =

Total amperage =
Total voltage drop =

Figure 17-11

3. Solve the series-parallel circuits in Figure 17-12 through Figure 17-14.

   a. Describe how series-parallel circuits apply to automotive electrical systems.

   b. Why are most circuits of the series-parallel design? _____

   c. What are the advantages of series-parallel circuits compared to series circuits and parallel circuits?

INSTRUCTOR VERIFICATION:

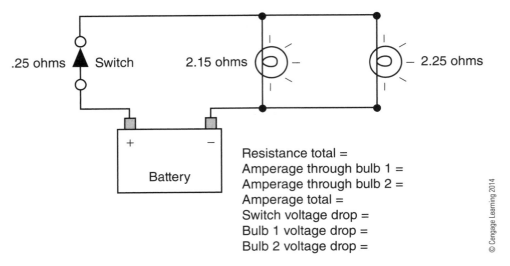

Resistance total =
Amperage through bulb 1 =
Amperage through bulb 2 =
Amperage total =
Switch voltage drop =
Bulb 1 voltage drop =
Bulb 2 voltage drop =

Figure 17-12

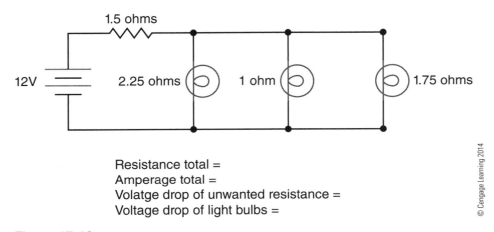

Resistance total =
Amperage total =
Volatge drop of unwanted resistance =
Voltage drop of light bulbs =

Figure 17-13

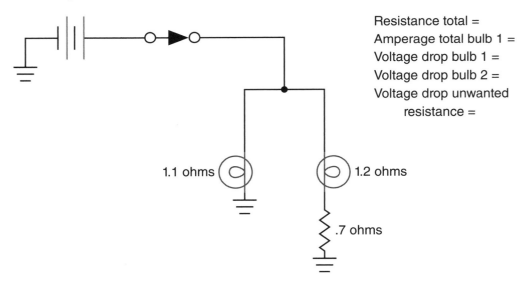

Resistance total =
Amperage total bulb 1 =
Voltage drop bulb 1 =
Voltage drop bulb 2 =
Voltage drop unwanted resistance =

Figure 17-14

INSTRUCTOR VERIFICATION:

4. Constructing an electromagnet

   A simple electromagnet can be built using a length of insulated wire and a nail or a screwdriver. Wrap a length of wire around a piece of metal, as shown in Figure 17-15. Connect the ends of the wire to a 9 V battery. The magnetic field produced should be strong enough to pick up small metal objects, such as paper clips.

   a. List at least five examples of automotive components that use electromagnetism.

   _____
   _____
   _____
   _____
   _____

   b. What do you think would result from using more voltage to power the electromagnet?

   _____
   _____

   c. How would increasing the size of the wire and the number of loops formed by the wire affect electromagnet strength?

   _____
   _____

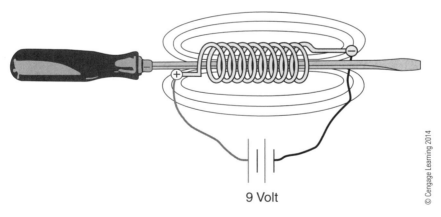

Figure 17-15

5. Label the types of circuit protection shown in Figure 17-16 (a through f).

   a. _____   b. _____
   c. _____   d. _____
   e. _____   f. _____

INSTRUCTOR VERIFICATION:

Chapter 17 Electrical/Electronic System Principles

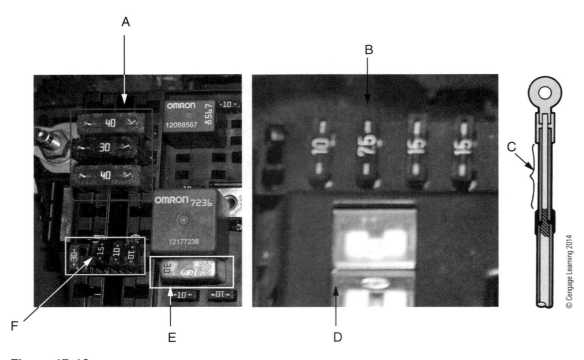

Figure 17-16

6. Refer to Figures 17-17 and 17-18 to complete the following table:

TABLE 17–1 Refer to Figures 17–17 and 17–18 and Complete the Following Table.

Fuse/Relay Location #	Fuse Size	Circuit Description	Fuse/Relay Type	Fuse Color
38				
		Radio		
	40			
6				
		Fuel pump	Micro relay	
		Spare		
25				
				Orange
			Diode	
12				

INSTRUCTOR VERIFICATION:

Chapter 17 Electrical/Electronic System Principles

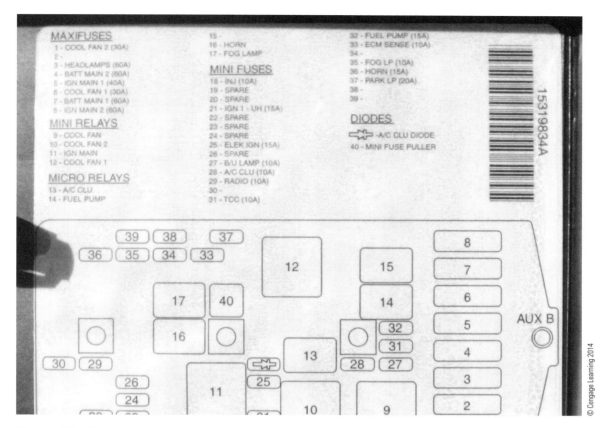

Figure 17-17

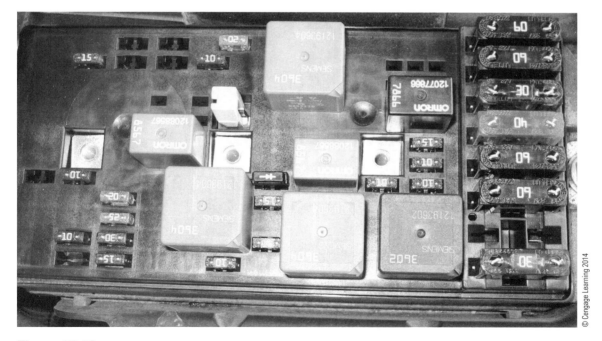

Figure 17-18

INSTRUCTOR VERIFICATION:

CHAPTER 18
Basic Electrical/Electronic System Service

Review Questions

1. When working on the electrical system, _____ should be your first priority.

2. Electrical shocks can also cause _____ due to the heat caused by current flow.

3. High-voltage wiring found in hybrid vehicles is which color?
   a. Orange
   b. Yellow
   c. Red
   d. Blue

4. List six safety precautions for working around automotive batteries.
   a. _____
   b. _____
   c. _____
   d. _____
   e. _____
   f. _____

5. A digital multimeter (DMM) is the primary tool to make _____, _____, and resistance measurements.

6. A _____ _____ can be used to check for power, ground, current flow, and continuity.

7. A current _____ is used to make inductive amperage measurements.

331

8. Describe electrostatic discharge and why it is dangerous for electronic components. _____
_____
_____
_____

9. Voltage is electrical _____ and is measured as the difference in potential between _____ points.

10. List the steps to perform a voltage measurement. _____
_____
_____
_____
_____

11. Why do DMMs have high internal resistance? _____
_____
_____
_____

12. How would using a DMM with low internal resistance affect a circuit when used to test voltage? _____
_____
_____

13. If using a DMM with manual voltage selections, which of the following settings would be used to measure vehicle battery voltage?
    a. 400 mv scale
    b. 4 volt scale
    c. 40 volt scale
    d. 400 volt scale

14. Define the term voltage drop. _____
_____
_____

15. Technician A says a voltage drop test can indicate excessive resistance in a circuit. Technician B says a voltage drop can be used to diagnose poor connections. Who is correct?
    a. Technician A
    b. Technician B
    c. Both A and B
    d. Neither A nor B

16. Voltage drop testing is a _____ test, meaning that the circuit must be in operation.

17. A voltage drop reading of 0.000 volts was obtained while testing a body ground. Technician A says the reading indicates a very good ground condition is present. Technician B says the test may not have been performed correctly. Who is correct?
    a. Technician A
    b. Technician B
    c. Both A and B
    d. Neither A nor B

18. The small amount of movement between terminals that can lead to wear and a resistive connection is called:
    a. Thermal cycling.
    b. Fretting.
    c. Voltage drop.
    d. High resistance.

19. To measure the resistance of a component or circuit, the power must be _____ before taking the measurement.

20. On a DMM, the Ω symbol is used to designate:
    a. Current flow.
    b. Voltage.
    c. Resistance.
    d. None of the above.

21. Explain what is meant by continuity and why you would test it. _____
    _____
    _____
    _____

22. A resistance test is a _____ test, meaning the circuit is not operating.

23. Explain why a resistance test will not always find the cause of an excessive voltage drop or decreased current flow. _____

_____

_____

_____

24. Current flow in a circuit is directly _____ to the resistance of the circuit.

25. While discussing measuring amperage with a DMM: Technician A says incorrect DMM setup can cause incorrect amperage readings to be displayed by the meter. Technician B says incorrect DMM setup can cause the fuse in the meter to blow. Who is correct?

    a. Technician A
    b. Technician B
    c. Both A and B
    d. Neither A nor B

26. When testing current flow with a DMM, the meter must be placed:

    a. Parallel to the circuit.
    b. In series with the circuit.
    c. In series-parallel with the circuit.
    d. None of the above.

27. To measure current flow above 10 amps, an _____ _____ clamp is used.

28. _____ are used to represent components in electrical wiring diagrams.

29. In a wiring diagram, a component drawn in a dashed line indicates:

    a. The complete component is shown.
    b. Not all of the component is shown.
    c. The component is not part of the circuit.
    d. The component is not accessible for testing.

30. Describe how color can be used to help the technician to understand wiring diagrams. _____

_____

_____

_____

31. Explain how to break a wiring diagram down into sections to understand circuit operation and how it can be tested. _____

_____

_____

_____

Chapter 18 Basic Electrical/Electronic System Service

32. Explain the three types of circuit faults and how each affects circuit operation.
_____
_____
_____
_____
_____

33. A vehicle has one headlight that is much brighter than the other headlight. Which type of circuit fault is most likely the cause?
    a. An open circuit
    b. A short to power
    c. A short to ground
    d. High resistance

34. An open circuit will cause a component to:
    a. Draw excessive amperage.
    b. Draw excessive voltage.
    c. Draw no current.
    d. Have excessive resistance.

35. A _____ circuit can be either to power or to ground.

36. Explain why a short to ground can be dangerous. _____
_____
_____

37. Before you begin to try to diagnose a short to ground, check to see if any aftermarket _____ have been recently installed in the vehicle.

38. Explain how to diagnose a short to ground. _____
_____
_____
_____

39. High resistance in a circuit can cause all of the following except:
    a. Excessive current flow.
    b. Dim lights.
    c. Slow motor operation.
    d. Inoperative components.

40. High resistance can also cause connectors to _____ and melt.

41. Unwanted resistance in the circuit causes a _____-than-normal voltage drop on the actual load in the circuit.

42. A _____ pack or splice connector is used to connect several circuits together.

43. The first step when testing a fuse is _____ the fuse.

44. What will be required if a fuse keeps blowing? _____

45. Describe how a circuit breaker functions. _____

46. Describe how to repair a section of broken wire. _____

47. Do not force _____ into the terminals while you are performing tests as this can spread the terminal apart or cause other damage, resulting in open circuits or intermittent connections.

48. Relays typically have a coil resistance of about:
    a. 10–20 ohms.
    b. 25–75 ohms.
    c. 50–150 ohms.
    d. 100–200 ohms.

49. Explain how to load-test a relay. _____

Activities

1. Label the electrical tools shown in Figure 18-1 (a through i).

   a. _____   b. _____
   c. _____   d. _____
   e. _____   f. _____
   g. _____   h. _____
   i. _____

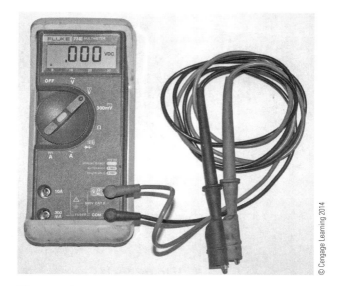

Figure 18-1a

Figure 18-1b

338 Chapter 18 Basic Electrical/Electronic System Service

Figure 18-1c

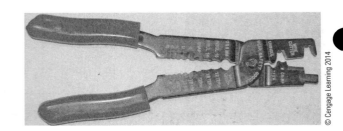

Figure 18-1d

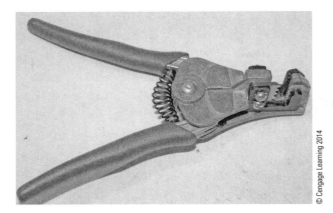

Figure 18-1e

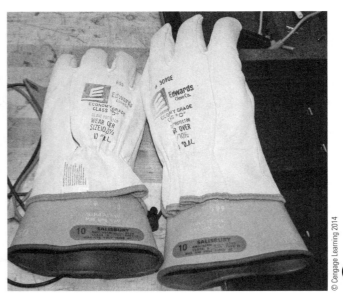

Figure 18-1f

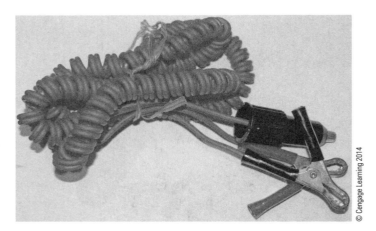

Figure 18-1g

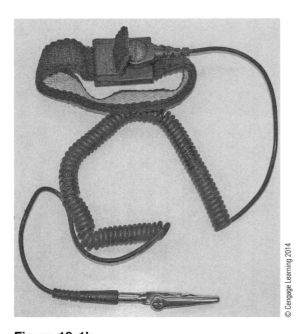

Figure 18-1h

Figure 18-1i

2. Label the functions of the meter shown in Figure 18-2 (a through h).

   a. _____   b. _____
   c. _____   d. _____
   e. _____   f. _____
   g. _____

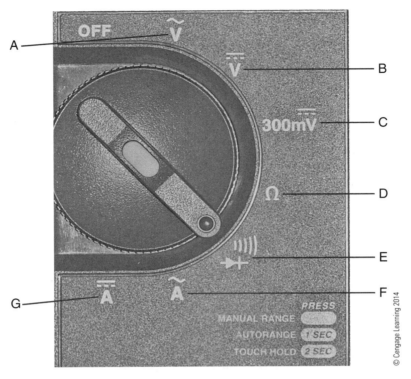

Figure 18-2

3. Label the types of tests shown in Figure 18-3 (a through c).

   a. _____   b. _____

   c. _____

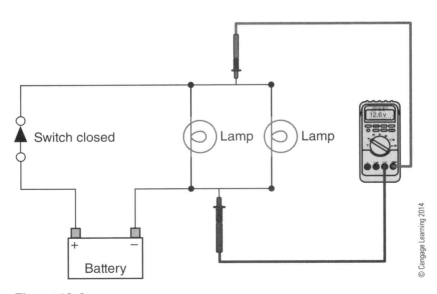

Figure 18-3a

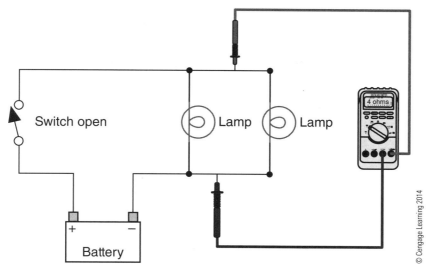

Figure 18-3b

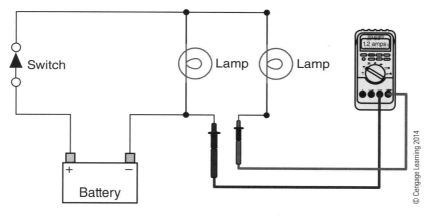

Figure 18-3c

4. Label the schematic symbols shown in Figure 18-4.

   a. _____   b. _____
   c. _____   d. _____
   e. _____   f. _____
   g. _____   h. _____
   i. _____   j. _____
   k. _____   l. _____
   m. _____

342 Chapter 18 Basic Electrical/Electronic System Service

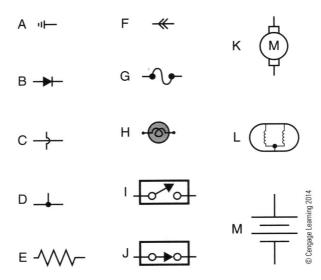

Figure 18-4

5. Label the terminals of the relays shown in Figure 18-5.

   a. _____   b. _____

   c. _____   d. _____

   e. _____   f. _____

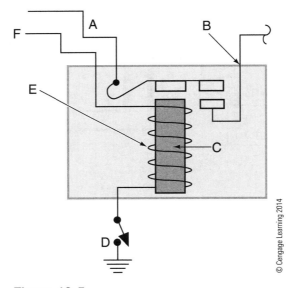

Figure 18-5

6. Match the following blade-type fuse ratings to the fuse colors.

3 amp	Green
5 amp	Tan/brown
10 amp	Yellow
15 amp	Violet
20 amp	Red
25 amp	Blue
30 amp	Clear

Lab Worksheet 18-1

Name ______________________ Date ______________ Instructor ______________

Using a DMM

Obtain from your instructor several automotive exterior lights with sockets. Bulb types 194, 1156, and 1157 (or similar) work well. Construct circuits using the diagrams below and record your findings.

1. **(Figure 18–6)** Series circuit.

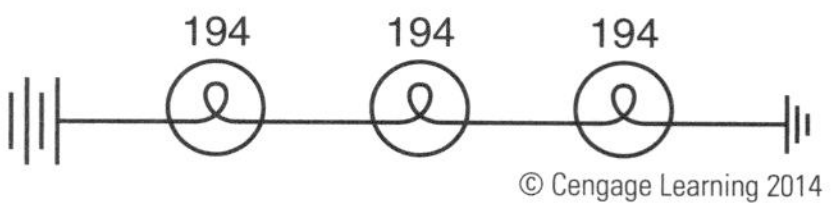

Figure 18-6 Series circuit.

Battery voltage: ______________ Circuit resistance: ______________

Voltage drop 1: ______________ Voltage drop 2: ______________ Voltage drop 3: ______________

Ohm's law predicted amperage: ______________ Measured amperage: ______________

2. **(Figure 18–7)** Series circuit.

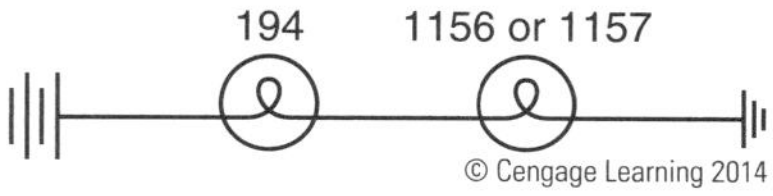

Figure 18-7 Parallel circuit.

Battery voltage: ______________ Circuit resistance: ______________

Voltage drop 1: ______________ Voltage drop 2: ______________

Ohm's law predicted amperage: ______________ Measured amperage: ______________

3. **(Figure 18–8)** Parallel circuit.

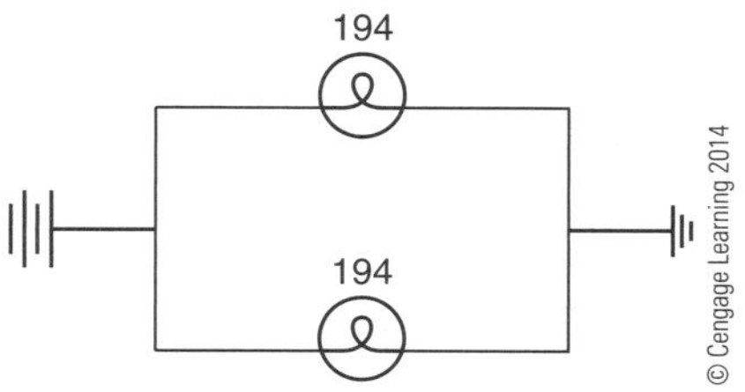

Figure 18-8 Parallel circuit.

Battery voltage: ______________ Circuit resistance: ______________

Voltage drop 1: ______________ Voltage drop 2: ______________

Ohm's law predicted amperage: ______________ Measured amperage: ______________

INSTRUCTOR VERIFICATION:

© 2014 Cengage Learning. All Rights Reserved. May not be scanned, copied or duplicated, or posted to a publicly accessible web site, in whole or in part.

4. **(Figure 18–9)** Parallel circuit.

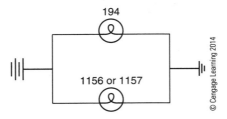

Figure 18-9 Parallel circuit.

Battery voltage: _____ Circuit resistance: _____

Voltage drop 1: _____ Voltage drop 2: _____

Ohm's law predicted amperage: _____ Measured amperage: _____

5. **(Figure 18–10)** Series-parallel circuit.

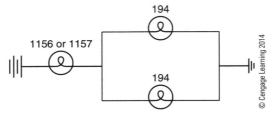

Figure 18-10 Series-parallel circuit.

Battery voltage: _____ Circuit resistance: _____

Voltage drop 1: _____ Voltage drop 2: _____

Voltage drop 3: _____ Voltage drop 4: _____

Ohm's law predicted amperage: _____ Measured amperage: _____

6. **(Figure 18–11)** Series-parallel circuit.

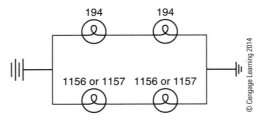

Figure 18-11 Series-parallel circuit.

Battery voltage: _____ Circuit resistance: _____

Voltage drop 1: _____ Voltage drop 2: _____ Voltage drop 3: _____

Voltage drop 4: _____ Voltage drop 5: _____ Voltage drop 6: _____

Ohm's law predicted amperage: _____ Measured amperage: _____

INSTRUCTOR VERIFICATION:

7. **(Figure 18–12)** Series-parallel circuit.

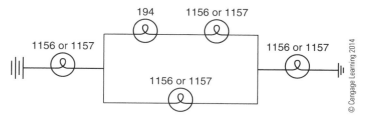

Figure 18-12 Series-parallel circuit.

Battery voltage: _____ Circuit resistance: _____ Voltage drop 1: _____

Voltage drop 2: _____ Voltage drop 3: _____ Voltage drop 4: _____

Voltage drop 5: _____ Voltage drop 6: _____ Voltage drop 7: _____

Ohm's law predicted amperage: _____ Measured amperage: _____

INSTRUCTOR VERIFICATION:

Lab Worksheet 18-2

Name _____ Date _____ Instructor _____

Using a Wiring Diagram

1. Examine the wiring diagram in Figure 18-13. The diagram represents a real vehicle you are testing. The circuit is for the radiator cooling fan with the fan operating normally—there are no problems with the circuit or the vehicle. Based on the diagram and normal operating conditions, record what a DMM would measure with the positive lead placed at the locations specified and the negative lead on battery ground.

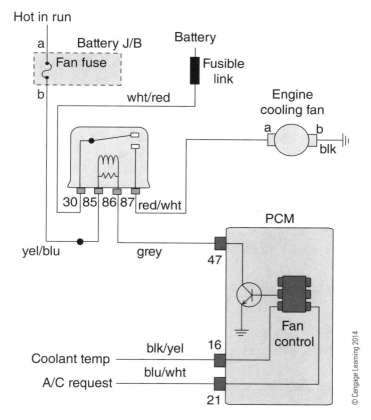

Figure 18-13

Remember that the fan is running during this testing.

Test Location	Expected DMM Reading
1. Terminal b at cooling fan motor	
2. Terminal 86 at cooling fan relay	
3. Terminal 30 at cooling fan relay	
4. Terminal 87 at cooling fan relay	
5. Terminal b at battery j/b	
6. Terminal 86 at cooling fan relay	
7. Terminal 47 at PCM	

INSTRUCTOR VERIFICATION:

Lab Worksheet 18-3

Name _____ Date _____ Instructor _____

Relay manufacturer _____ Part number _____ Number of pins _____

Relay Testing

1. Describe the external condition of the relay. _____
   _____

2. Draw the pin configuration and label the terminals as shown on the relay.

   If the relay terminals are not labeled, draw the pin configuration and label the terminals after testing.

3. Using a DMM, measure and record the relay coil resistance. _____

   Is the resistance within specs?   Yes _____   No _____   Unknown _____

4. Apply power and ground to the coil terminals. Does the relay click?

   Yes _____   No _____

   If no, what may this indicate? _____

5. With the relay coil energized, measure the resistance across the two remaining switch terminals. Resistance measurement _____

   What is indicated by this measurement? _____

6. With the relay coil energized, connect a power source and a test light to the switch terminals. Does the test light illuminate?   Yes _____   No _____

7. Based on your tests, what is the condition of the relay? _____
   _____

INSTRUCTOR VERIFICATION:

Starting and Charging System Principles

Review Questions

1. The parts of a battery are the ___Anode___, the ___cathode___, and the electrolyte.

2. The electrolyte in an automotive battery is made of ___Sulfuric___ acid and ___Water___.

3. A fully charged automotive battery produces how much voltage per cell?
   a. 2 volts
   (b.) 2.1 volts
   c. 2.2 volts
   d. 12.6 volts

4. As an automotive battery discharges, the electrolyte becomes mostly ___Water___.

5. The main function of the automotive battery is to power the ___starting___ motor.

6. List and explain the four stages of battery operation. ___charged, discharging, discharged,___ ___charging___

7. Over time, the discharge and recharge process causes the battery to lose ___electrolytes___.

8. Which of the following affect how well a battery accepts a charge?
   a. Battery temperature
   b. Battery state of charge
   c. Plate sulfation
   (d.) All of the above

Chapter 19 Starting and Charging System Principles

9. Technician A says overcharging a battery can shorten the battery life. Technician B says excessive vibration can shorten battery life. Who is correct?
   a. Technician A
   b. Technician B
   c. Both A and B ✓
   d. Neither A nor B

10. Which type of automotive battery may require water to be added occasionally?
    a. Maintenance-free battery
    b. AGM battery
    c. Low-maintenance battery ✓
    d. Deep-cycle battery ✓

11. All of the following statements about hybrid vehicles are correct except:
    a. Some hybrid vehicles have both high-voltage and low-voltage batteries. ✓
    b. The high-voltage battery is a different design than a lead-acid battery.
    c. The high-voltage battery is not easily accessible.
    d. A high-voltage system disconnect is used to isolate the high-voltage battery.

12. The high-voltage wiring found on hybrids is covered in bright __Orange__ conduit.

13. The cold cranking amps rating is based on battery power available at __0__ degrees F.

14. The cranking amps rating is based on battery power available at __32°__ degrees F.

15. The battery reserve capacity rating is used to rate in __Minutes__ battery output if the charging system fails.

16. Which battery rating is used to define a battery's size and terminal type?
    a. CCA
    b. A-h
    c. CA
    d. BCI ✓

17. The BCI group rating defines all of the following battery characteristics except:
    a. Terminal position.
    b. Battery height.
    c. CCA rating. ✓
    d. Overall battery size.

Chapter 19 Starting and Charging System Principles 353

18. List eight examples of battery handling and service safety precautions. _____
_____
_____
_____
_____
_____
_____
_____

19. The function of the ___Starter motor___ is to spin the engine fast enough that the air-fuel mixture can be compressed and ignited so that combustion can begin.

20. When current flows through a conductor, a ___electro magnetic___ field is produced around the conductor.

21. The parts of the electric motor responsible for switching the polarity of the magnetic fields are the:
    a. Armature and permanent magnets.
    b. Permanent magnets and commutator.
    c. Armature and brushes.
    d. Brushes and commutator.

22. Which of the following is the rotating part of the starter motor?
    a. Brushes
    b. Permanent magnets
    c. Armature
    d. Field coils

23. The solenoid completes the circuit to the starter motor and is used to pull a _____ that kicks out the drive gear into the flywheel.

24. Label the starter connections shown in Figure 19-1.
    a.
    b.
    c.

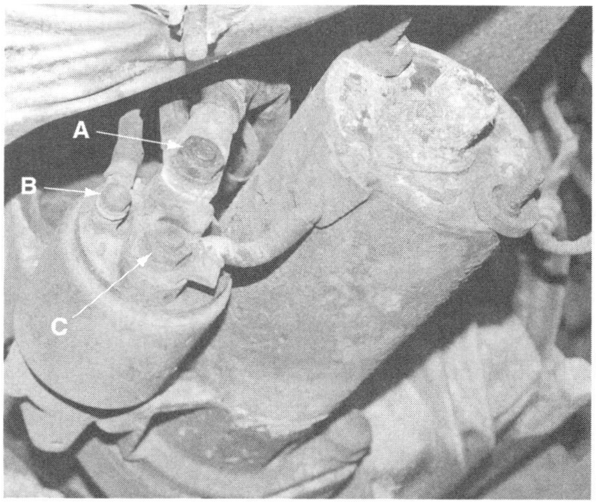

Figure 19-1

Chapter 19 Starting and Charging System Principles

25. List the components of a typical starting system. _____
_____
_____

26. The three starter circuits are the _____ power, _____ ground, and the _____ circuit.

27. Explain the purpose of the park/neutral switch in the starting system. _____
_____
_____

28. Some antitheft systems use a starter _____ relay to prevent or allow starter operation based on the theft system status.

29. In newer vehicles, the ignition switch often does not supply power directly to the _____ circuit; instead, the switch is used as part of the control circuit that includes starter relays and _____ modules.

30. Vehicles that have Push to Start also use a _____ key to validate the starting process.

31. Technician A says full hybrids use a conventional 12-volt starter motor to crank the engine. Technician B says full hybrids use a high-voltage motor/generator to crank the engine. Who is correct?
    a. Technician A
    b. Technician B
    c. Both A and B
    d. Neither A nor B

32. True or False: Some nonhybrid vehicles use an engine idle stop-start function to improve fuel economy.
_____

33. Once the engine has started, the _____ _____ takes over powering the vehicle's electrical system.

34. The AC _____ is used to recharge the battery once the engine is started.

35. A modern generator produces as much as:
    a. 30 amps.
    b. 50 amps.
    c. 80 amps.
    d. 140 amps.

Chapter 19   Starting and Charging System Principles   355

36. List the components of the generator rotor. _____
_____
_____

37. What is the purpose of the AOD or overrunning clutch? _____
_____
_____

38. The stator typically has _____ windings of wire.

39. _____ are used in the generator to convert AC into DC.

40. Which component controls the flow of current to the rotor coil?
    a. Stator
    b. Slip rings
    c. Voltage regulator
    d. Pole pieces

41. Explain the four factors of induction. _____
_____
_____
_____
_____

42. Technician A says generator field control is always on the ground side of the circuit. Technician B says voltage regulators may control the power or the ground side of the field. Who is correct?
    a. Technician A
    b. Technician B
    c. Both A and B
    d. Neither A nor B

43. Voltage regulators may be located:
    a. Internal to the generator.
    b. External to the generator.
    c. In the PCM.
    d. All of the above.

44. Describe how the PCM can be used to control generator output. _____
_____
_____
_____

Activities

1. Match the battery cycle with the proper label in Figure 19-2 (a through d).

Charging   Discharged   Charged   Discharging

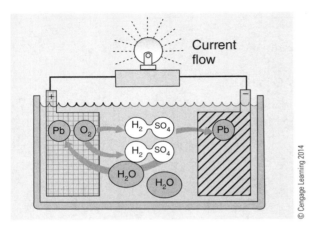

Figure 19-2a

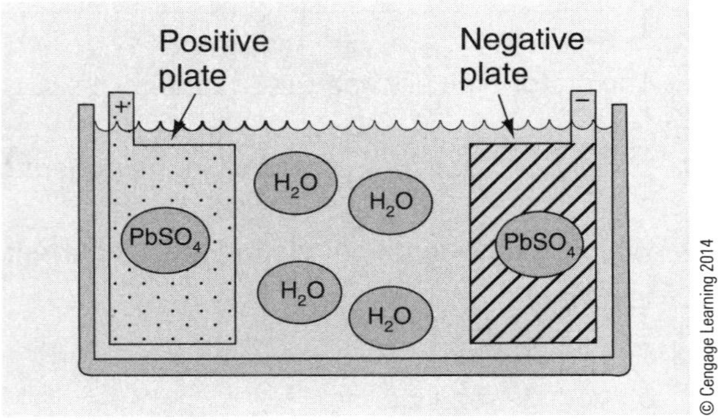

Figure 19-2b

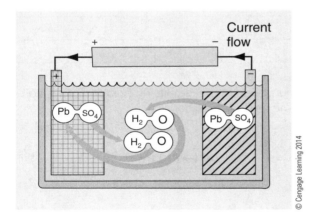

Figure 19-2c

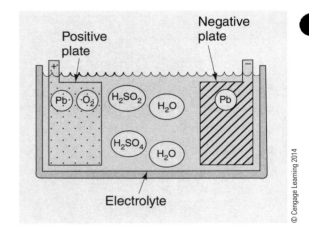

Figure 19-2d

2. Label the parts of the battery shown in Figure 19-3.

a. ____________________ b. ____________________

c. ____________________ d. ____________________

e. ____________________ f. ____________________

© 2014 Cengage Learning. All Rights Reserved. May not be scanned, copied or duplicated, or posted to a publicly accessible web site, in whole or in part.

Chapter 19 Starting and Charging System Principles 357

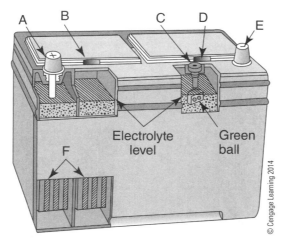

Figure 19-3

3. Label the parts of the starters in Figure 19-4 (a through c).

   a. _____   b. _____
   c. _____   d. _____
   e. _____   f. _____
   g. _____   h. _____
   i. _____

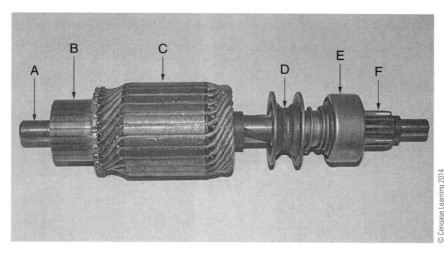

Figure 19-4a

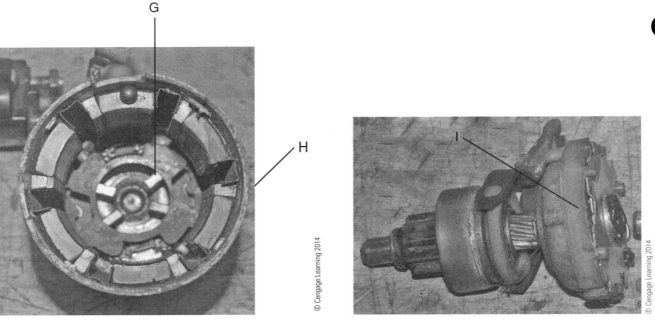

Figure 19-4b

Figure 19-4c

4. Obtain from your instructor two transmission pan type magnets, a two- to three-foot length of insulated single-strand wire like that used for armature windings on a fan motor, paper clips, or cotter pins, a Styrofoam cup or small box, and a 9 V battery. Follow these steps to build a simple DC electric motor:

   - Strip the ends of the wire to remove about one-half of the insulation along the length of the last two inches of the wire. Each end should have bare copper exposed around half of the wire's circumference and insulation around the other half. This will allow an on/off connection through the wire.
   - Form the wire (coil) into a tightly wound loop, approximately one inch in diameter, leaving the two ends protruding out from the middle, as shown in Figure 19-5.

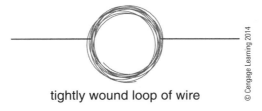

tightly wound loop of wire

Figure 19-5

   - Mount the paper clips or cotter pins into the cup or box so that they can form a holding fixture for the wire.
   - Place the magnets so that they hold each other to the top section of the cup or box between the paper clips or cotter pins.
   - Place the coil into the paper clips or cotter pins. Adjust the height so that the coil can rotate very close to the magnet without touching.
   - Attach the 9 V battery so that the positive is connected to one paper clip or cotter pin and the negative is connected to the other paper clip or cotter pin, as shown in Figure 19-6.
   - Some adjusting may be required, but once a good connection is made, the coil should easily spin on its own.

INSTRUCTOR VERIFICATION:

Chapter 19 Starting and Charging System Principles 359

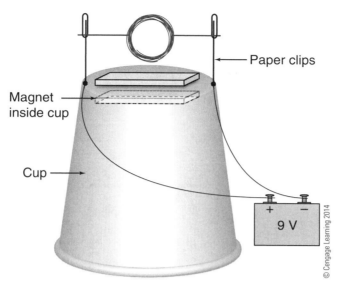

Figure 19-6

a. Note the coil's direction of rotation. _____

b. Reverse the battery polarity and note the motor rotation. _____

c. Why did the motor respond as it did to the battery voltage change in polarity? _____

d. List the electrical motors used in modern automobiles that operate so that they reverse their direction of rotation. _____

e. List the electrical motors used in modern automobiles that rotate in only one direction. _____

f. What would be the result if a motor, such as that for the cooling fan which only rotates one way, were hooked to the battery in reverse polarity? _____

g. Match the motor components with their electrical functions.

Paperclip/cotter pin	Armature
Wire	Brush
Magnets	Commutator
Wire insulation	Field coils

h. Explain why the wire's insulation is not completely removed. _____

i. What effect would using a smaller battery, such as a D cell, have on the motor and why? _____

j. What effect would using more magnets have on the motor and why? _____

k. Why do you think the electric motors used in automobiles have more commutator segments than the simple motor you built? _____

l. What are the wear points on a DC motor? _____

m. What are some examples of problems that can affect the operation of a DC motor? _____

INSTRUCTOR VERIFICATION:

5. Label the parts of the starting circuit shown in Figure 19-7.
   a. _____   b. _____
   c. _____   d. _____
   e. _____   f. _____
   g. _____

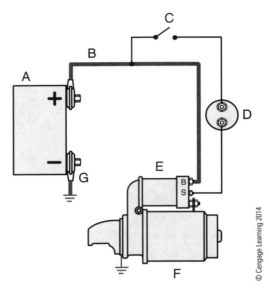

Figure 19-7

6. Label the parts of the charging system shown in Figure 19-8.
   a. _____   b. _____
   c. _____   d. _____
   e. _____   f. _____
   g. _____

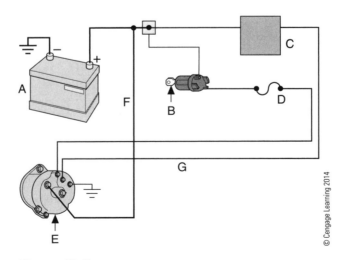

Figure 19-8

INSTRUCTOR VERIFICATION:

Lab Worksheet 19-1

Name _____ Date _____ Instructor _____

Year _____ Make _____ Model _____

Identify Starting and Charging Components

1. Inspect the battery and note the following:

   CCA rating _____   CA rating _____

   RC rating _____   A-h rating _____

   BCI group number _____

2. Note the location of the battery. _____

3. Locate the starter motor and note its location. _____

4. Using service information, determine the starter type.

   Direct drive _____   Gear reduction

5. Locate the AC generator and note its location. _____

6. Look on the generator or in the service information for the generator's rated output.

   _____ Amps   _____ Volts

7. Using the service information, determine how the generator field is controlled.

   _____
   _____

INSTRUCTOR VERIFICATION: _____

CHAPTER 20

Starting and Charging System Service

Review Questions

1. Describe four tools used in diagnosing the starting and charging systems.

   _____
   _____
   _____
   _____

2. Describe three tools used for servicing batteries. _____
   _____
   _____
   _____

3. List six safety precautions for working around batteries.
   a. _____
   b. _____
   c. _____
   d. _____
   e. _____
   f. _____

4. A starter should never be cranked for more than _____ seconds at a time without at least a _____ minute cool-off period.

5. HEV high-voltage wiring and components are identified by the _____ colored conduit and connections.

6. Technician A says all modern vehicles place the 12V battery under the hood. Technician B says some 12V batteries are located inside the passenger compartment. Who is correct?
   a. Technician A
   b. Technician B
   c. Both A and B
   d. Neither A nor B

7. Battery corrosion can be neutralized by using _____ and water as a paste.

8. List five types of problems the battery case should be inspected for. _____

9. Technician A says all lead-acid batteries have removable cell caps so that the electrolyte levels can be checked. Technician B says only maintenance-free batteries have removable cell caps. Who is correct?
   a. Technician A
   b. Technician B
   c. Both A and B
   d. Neither A nor B

10. If the terminal is loose on the post, an _____ connection problem can occur.

11. Side-post batteries can be _____ by overtightening the side-post connections.

12. List three things battery cables should be inspected for.
    a. _____
    b. _____
    c. _____

13. A loose battery from a missing _____ can be damaged by bouncing around.

14. Many batteries have a built-in _____ to indicate the battery state of charge.

15. Explain what is meant by a battery voltage leak and how to test for it. _____

16. A battery is low on electrolyte: Technician A says the battery can be topped off with tap water. Technician B says the generator may be undercharging the battery, causing the low acid level. Who is correct?
    a. Technician A
    b. Technician B
    c. Both A and B
    d. Neither A nor B

17. A fully charged battery should have at least _____ volts.
    a. 12.6
    b. 12.2
    c. 12.9
    d. 12.0

18. While discussing battery charging: Technician A says the best way to recharge a battery is with a high charge rate over a long period of time. Technician B says a low charge rate over several hours is less likely to damage the battery. Who is correct?
    a. Technician A
    b. Technician B
    c. Both A and B
    d. Neither A nor B

19. List the steps to properly connect and set a battery charger. _____
    _____
    _____
    _____

20. Which type of automotive battery requires special charging procedures?
    a. Maintenance-free batteries
    b. AGM batteries
    c. Low-maintenance batteries
    d. None of the above

21. A key-off drain, also called a _____ load, continues to draw power from the battery after the vehicle is shut off.

22. Which of the following methods can be used to test key-off battery drain?
    a. Inductive current clamp
    b. Test light in series with the battery and battery cable
    c. DMM in series with the battery and battery cable
    d. All of the above

23. Key-off battery drain should generally be less than:
    a. 200ma
    b. 100ma
    c. 50ma
    d. 10ma

24. The three-minute charge test is used to determine if the battery plates are _____.

25. Describe how to perform a battery load test. _____

26. If the battery load capacity is unknown, explain another method to determine how much load to apply during a battery load test. _____

27. A battery load test is performed for _____ seconds.

28. Explain why it is important to use a memory-saving device when disconnecting a battery for service.

29. List the proper order in which to connect two batteries for jump starting. _____

30. Describe the procedure to disable the high-voltage system on a hybrid vehicle.

31. Describe how to voltage drop test the starter control circuit. _____

32. The starting system in Figure 20-1 is being tested for a no-crank condition. With the key in the start position, the voltage reading shown in Figure 20-1 is observed. Technician A says this indicates the neutral safety switch is open. Technician B says the solenoid may be faulty. Who is correct?

   a. Technician A
   b. Technician B
   c. Both A and B
   d. Neither A nor B

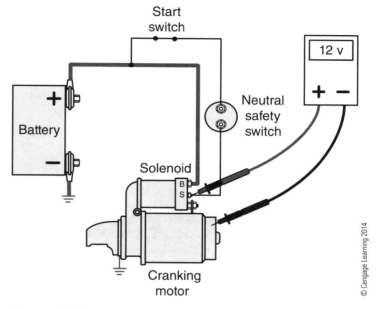

Figure 20-1

33. The circuit shown in Figure 20-2 is being tested for a no-crank condition. Based on the voltage reading shown, which is the mostly likely cause?

   a. Faulty starter motor
   b. Faulty neutral safety switch
   c. Faulty ignition switch
   d. Discharged battery

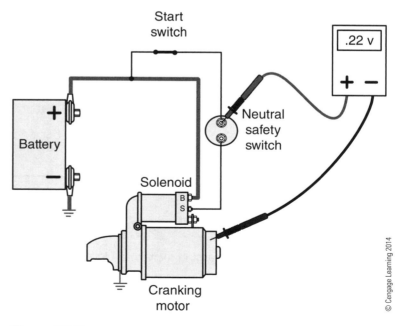

Figure 20-2

Chapter 20 Starting and Charging System Service

34. A vehicle's starter clicks rapidly when the key is turned to the start position: which is the most likely cause?
    a. Low battery state of charge
    b. Defective solenoid
    c. Shorted starter windings
    d. Defective park-neutral switch

35. Before attempting to test the starter, ensure that the _____ is charged and capable of producing the necessary _____ required by the starter.

36. Which of the following can cause lower-than-normal starter motor current draw?
    a. Shorted starter field coil windings
    b. Binding of the armature
    c. Low engine compression
    d. Open solenoid windings

37. Technician A says starter control circuit voltage drop should be less than 200mv. Technician B says starter motor circuit voltage drop should be less than 500mv. Who is correct?
    a. Technician A
    b. Technician B
    c. Both A and B
    d. Neither A nor B

38. If the voltage drops of the battery cables are excessive, remove and clean the battery cable connections at the _____ and the _____.

39. Before removing the starter motor from the vehicle, first disconnect the _____.

40. When removing a starter, do not let the starter motor hang by the _____ or control circuit _____.

41. Some starters require a _____ to set the clearance between the starter and the flywheel.

42. Explain why some starters are bench tested for current draw instead of being tested on the vehicle.

_____

_____

_____

43. Begin the inspection of the charging system by looking at the _____ indicator light on the instrument panel.

44. List five things to check the generator drive belt for.

   a. _____
   b. _____
   c. _____
   d. _____
   e. _____

45. A loose generator drive belt can cause:
   a. An overcharge condition.
   b. An undercharge condition.
   c. A discharged battery.
   d. Both b and c.

46. A _____ battery can affect the generator's performance.

47. Technician A says for the charging system to operate properly, the battery must be in good condition. Technician B says battery condition should not affect generator operation. Who is correct?
   a. Technician A
   b. Technician B
   c. Both A and B
   d. Neither A nor B

48. A properly operating charging system should be able to produce at least _____ of the generator's rated output.
   a. 50 percent
   b. 60 percent
   c. 80 percent
   d. 90 percent

49. Describe how to test for AC voltage leaking from the generator. _____
_____
_____
_____

50. List five problems that can cause an undercharge condition.

   a. _____
   b. _____
   c. _____
   d. _____
   e. _____

51. An open field winding will cause which of the following concerns?
    a. Overcharging
    b. Undercharging
    c. No charging
    d. None of the above

52. When testing generator output, you may need a _____ to check for DTCs and data for the charging system.

53. An _____ problem occurs when the field control circuit does not limit the amount of current that is supplied to the field.

54. When removing a generator, first disconnect the _____ to prevent damage to the electrical system.

Activities

1. A vehicle has a no-crank condition: When the headlights are turned on and the ignition turned to start, no noise is heard and the lights do not dim. Which of the following should be checked first and why?

   Battery connections   Ignition switch

   Battery voltage   Drive belt

   Starter connections   Generator connections

   Starter solenoid   Starter motor

2. Place the connections shown in Figure 20-3 in the correct order to jump-start a vehicle.

   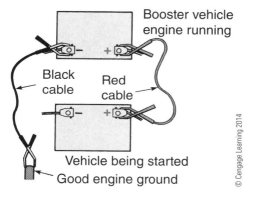

   Figure 20-3

3. Identify the tests shown in Figure 20-4 (a through d).
   a. _____
   b. _____
   c. _____
   d. _____

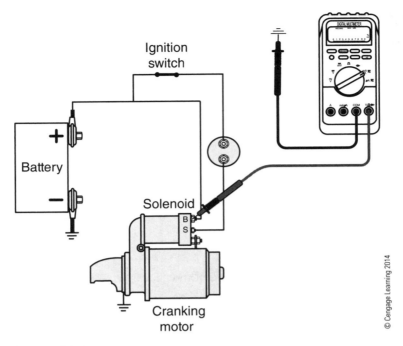

Figure 20-4a

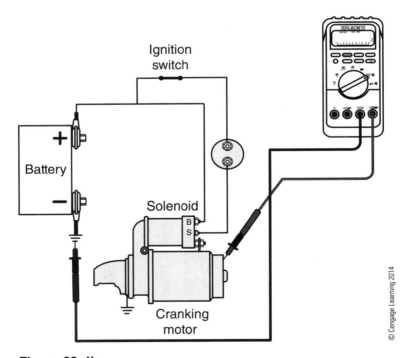

Figure 20-4b

Chapter 20 Starting and Charging System Service 373

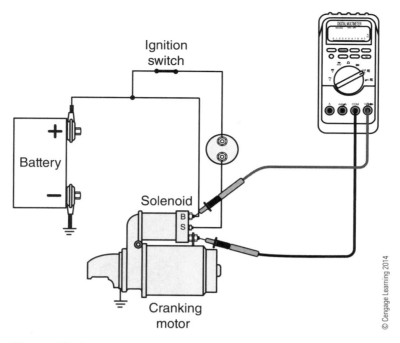

Figure 20-4c

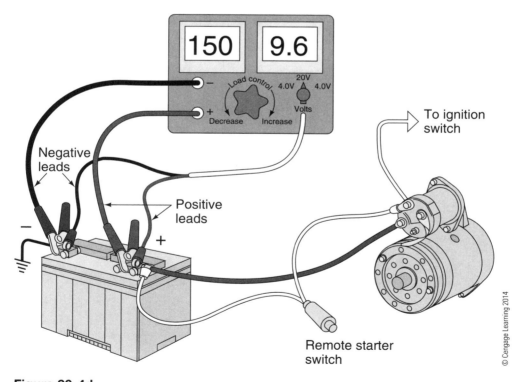

Figure 20-4d

Lab Worksheet 20-1

Name _____ Date _____ Instructor _____

Year _____ Make _____ Model _____

Battery group number _____ RC _____

CCA _____ CA _____ AH _____

Battery Inspection

1. Inspect and note the battery hold-down condition. _____

2. Inspect the case for damage and acid leaks.   Damage/leaks   Yes _____   No _____

3. Battery open circuit voltage _____

   What is the battery state of charge based on the open circuit voltage? _____

4. Inspect the terminals for fit, damage, and corrosion. Note your findings. _____
   _____

5. Inspect the cables for the following concerns:
   a. Damaged insulation   Yes _____   No _____
   b. Broken strands   Yes _____   No _____
   c. Secure connections   Yes _____   No _____

6. If the battery has removable cell caps, remove the caps and note the acid level in each cell.

   Cell 1 _____ Cell 2 _____ Cell 3 _____ Cell 4 _____ Cell 5 _____ Cell 6 _____

   a. What may be indicated by low acid level in one or more cells? _____
   _____

7. Based on your inspection, what is the general condition of the battery? _____
   _____
   _____

INSTRUCTOR VERIFICATION:

Lab Worksheet 20-2

Name _____ Date _____ Instructor _____

Year _____ Make _____ Model _____

Battery group number _____ CCA/CA _____

Battery Open Circuit Voltage Test

1. Explain the open circuit voltage test. _____
   _____

2. Measure and record the open circuit voltage. _____

3. Is the voltage correct?   Yes _____   No _____
   a. If no, what can cause the voltage to be low? _____
      _____
   b. What can cause the voltage to be high? _____
      _____

4. Does the battery require charging?   Yes _____   No _____

5. Based on your test, what is the condition of the battery? _____
   _____

INSTRUCTOR VERIFICATION:

Lab Worksheet 20-3

Name _____ Date _____ Instructor _____

Year _____ Make _____ Model _____

Battery group number _____ CCA/CA _____

Conductance Testing a Battery

1. Explain the battery conductance test. _____
   _____

2. Tester used _____

3. Before testing, use a DMM to determine the battery state of charge.
   a. Open circuit voltage reading _____
   b. Is the battery sufficiently charged to continue?   Yes _____   No _____

4. Connect the conductance tester and program as necessary.

5. Perform the test and record the results. _____
   _____

6. Based on this test, what is the condition of the battery? _____
   _____

INSTRUCTOR VERIFICATION: _____

# Lab Worksheet 20-4

Name _____ Date _____ Instructor _____

Year _____ Make _____ Model _____

Battery group number _____ CCA/CA _____

Battery Capacity Test

1. Explain the battery capacity test. _____
   _____

2. How is the load test amount determined for this battery? _____
   _____

3. If the battery CCA is unknown, how can the load test amount be determined? _____
   _____

4. Battery load test amount _____ amps

5. Connect the tester to the battery and ensure the connections are secure.

6. Load-test the battery for 15 seconds.

   Voltage during test _____ Amperage during test _____

7. Let the battery stabilize for two minutes and record the voltage. _____ volts

8. Based on your testing, what is the condition of the battery? _____
   _____

INSTRUCTOR VERIFICATION: _____

Lab Worksheet 20-5

Name _____ Date _____ Instructor _____

Year _____ Make _____ Model _____

Battery group number _____ CCA/CA _____

Three-Minute Charge (Sulfation) Test

1. Explain what plate sulfation means and how it affects battery performance. _____
   _____
   _____

2. Ensure that the battery is a low-maintenance or maintenance-free battery and not an AGM battery.
   a. Why should this test not be performed on AGM batteries? _____
      _____

3. Record the battery open circuit voltage before charging. _____ volts

4. Connect a DMM to the battery terminals to monitor battery voltage. Voltage should not exceed 15.5 volts at the end of the three minutes.

5. Connect the battery charger and set the charge rate to 40 amps.

6. Record the battery voltage after three minutes. _____ volts

7. Based on this test, what is the condition of the battery? _____
   _____

INSTRUCTOR VERIFICATION: _____

Lab Worksheet 20-6

Name _____ Date _____ Instructor _____

Year _____ Make _____ Model _____

Battery group number _____ CCA/CA _____

Parasitic Draw Testing

1. Explain why a parasitic draw test is performed. _____
   _____

2. Check which of the three methods of testing parasitic draw will be used.
   _____ Test light
   _____ DMM
   _____ Inductive ammeter

3. Explain why using a test light is the least accurate parasitic draw test method. _____
   _____

4. Connect the test equipment and measure the parasitic draw.
   _____ amps

5. Locate the manufacturer's specification for allowable parasitic draw and record.
   _____ amps

6. Does the measured draw exceed the spec?   Yes _____   No _____

7. If the draw is excessive, what are the possible causes of the draw? _____
   _____

INSTRUCTOR VERIFICATION:

Chapter 20 Starting and Charging System Service 387

Lab Worksheet 20-7

Name _____ Date _____ Instructor _____

Year _____ Make _____ Model _____

Battery group number _____ CCA/CA _____

Starter Circuit Testing

1. Measure and record the battery voltage. _____ volts

   Connect the starter testing equipment and record cranking volts and amps.

2. Cranking volts _____   Cranking amps _____

3. Compare your readings with the manufacturer's specifications. Are the cranking volts and amps within specs?
   Yes _____   No _____

4. Connect a DMM to the battery positive terminal and the starter positive cable connection. Crank the engine and record the voltage.
   _____ volts   Is this voltage drop acceptable?   Yes _____   No _____

5. Connect a DMM to the starter case and the battery negative terminal connection. Crank the engine and record the voltage.
   _____ volts   Is this voltage drop acceptable?   Yes _____   No _____

6. Connect a DMM to the starter control circuit terminal at the solenoid and the battery positive terminal connection. Crank the engine and record the voltage.
   _____ volts   Is this voltage drop acceptable?   Yes _____   No _____

7. Based on your testing, what is the condition of the starter and starter circuit? _____
   _____

INSTRUCTOR VERIFICATION: _____

Lab Worksheet 20-8

Name _____ Date _____ Instructor _____

Year _____ Make _____ Model _____

Generator Drive Belt Inspection

1. Locate and note the type of drive belt used.

    V-belt _____   Serpentine (multirib) belt _____

2. Describe how tension is applied to the generator belt. _____
   _____

3. Inspect the drive belt for wear and damage. Note your findings. _____
   _____
   _____

4. Locate the belt tension specifications and record them.   Spec _____

5. Using a belt tension gauge, measure and record the generator drive belt tension.

    Measured tension _____

6. How does the measured tension compare to the specification? _____
   _____

7. Based on your inspection, what do you recommend? _____
   _____
   _____

INSTRUCTOR VERIFICATION:

Lab Worksheet 20-9

Name _____ Date _____ Instructor _____

Year _____ Make _____ Model _____

Generator Output Testing

1. Test battery voltage and record. _____ volts

2. Inspect the drive belt for wear, damage, and tension. Record your findings. _____
   _____

3. Visually inspect the wiring to the generator. Record your findings. _____
   _____

4. Locate and record the generator output rating. _____ amps

5. With a starting/charging system tester connected to the battery, record the charging output at idle with no loads turned on.
   _____ volts   _____ amps

6. With several loads turned on, record the idle output.
   _____ volts   _____ amps

7. Run the engine at 2,000 rpm and record the charging output with no loads.
   _____ volts   _____ amps

8. Load the generator until battery voltage reaches 12.6 V and record the output.
   _____ amps

9. Based on your testing, what is the condition of the charging system? _____
   _____

INSTRUCTOR VERIFICATION:

Lab Worksheet 20-10

Name _____ Date _____ Instructor _____

Year _____ Make _____ Model _____

Charging System Voltage Drop Testing

Connect a starting/charging system tester to the vehicle.

1. Measure the voltage drop between the battery positive and the generator output terminal with the engine at idle.
   _____ Voltage drop no load
   _____ Voltage drop with load

2. Measure the voltage drop between the battery positive and the generator output terminal with the engine at 2,000 rpm.
   _____ Voltage drop
   _____ Voltage drop with load

3. Measure the voltage drop between the battery negative and the generator case with the engine at idle.
   _____ Voltage drop
   _____ Voltage drop with load

4. Measure the voltage drop between the battery negative and the generator with the engine at 2,000 rpm.
   _____ Voltage drop
   _____ Voltage drop with load

5. Did the voltage drop change with a load applied?   Yes _____   No _____

6. If yes, why do you think there was a change in the voltage drop amounts? _____
   _____

7. Based on your testing, what is the condition of the charging system's wiring?
   _____

INSTRUCTOR VERIFICATION: _____

CHAPTER 21
Lighting and Electrical Accessories

Review Questions

1. Which of the following is not an advantage of LED lights?
   a. Increased bulb life
   b. Increased heat generation
   c. Decreased power consumption
   d. Decreased heat output

2. Door switches are used to either _____ a light circuit or to serve as an input for the _____.

3. Vehicles that have interior lights that slowly _____ and then go out use either the lighting or body _____ to control the lights.

4. Instrument panel illumination is controlled through the _____ switch.

5. When you are beginning the diagnosis of a customer's concern, begin by _____ the complaint.

6. Technician A says both incandescent and LED lights can be checked with an ohmmeter. Technician B says only incandescent lights can be checked with an ohmmeter. Who is correct?
   a. Technician A
   b. Technician B
   c. Both A and B
   d. Neither A nor B

7. A switch should show continuity when the switch is _____.

8. Explain why a more accurate test of a switch is performed by measuring its voltage drop during operation than by testing resistance only.

9. _____ means that there is a complete circuit or path between the meter's test leads.

10. In many vehicles, the door switches complete a _____ circuit through the switch itself.

Chapter 21 Lighting and Electrical Accessories

11. Why is it important to use care and caution when removing interior lighting components?

12. Explain bulb trade numbers. _____
_____
_____

13. None of the IP bulbs operate: Technician A says the IP dimmer control may be faulty. Technician B says the IP dimmer may be turned down. Who is correct?
    a. Technician A
    b. Technician B
    c. Both A and B
    d. Neither A nor B

14. All of the following are types of exterior lights except:
    a. Back-up lights.
    b. IP lights.
    c. Turn signal lights.
    d. Stoplights.

15. Halogen insert bulbs have replaced _____ headlights because of their size and improved design flexibility.

16. HID lights use _____-_____ AC to create an arc across two electrodes to produce light.

17. The _____ switch is used to switch between the low- and high-beam headlights.

18. DRL stands for _____.

19. Parking lamps are front, rear, or side marker bulbs operated by the _____ switch.

20. If the vehicle uses the same bulbs for the brake and _____, then the turn signal circuit connects to the brake light circuit.

21. Explain the operation of a bimetallic turn signal flasher unit. _____
_____
_____

22. Explain how using an incorrect wattage bulb can affect turn signal circuit operation.
_____
_____
_____

23. List four systems that often use the brake light switch as an input.

    a. _____
    b. _____
    c. _____
    d. _____

24. Define CHMSL. _____
    _____

25. The clear lights at the rear of the vehicle are used as:

    a. Brake lights.
    b. Turn signal lights.
    c. Back-up lights.
    d. All of the above.

26. A vehicle with the headlight system shown in Figure 21-1 has no low-beam light operation. Technician A says a poor connection at terminal 2 of the dimmer switch could be the cause. Technician B says an open ground at G102 could be the cause. Who is correct?

    a. Technician A
    b. Technician B
    c. Both A and B
    d. Neither A nor B

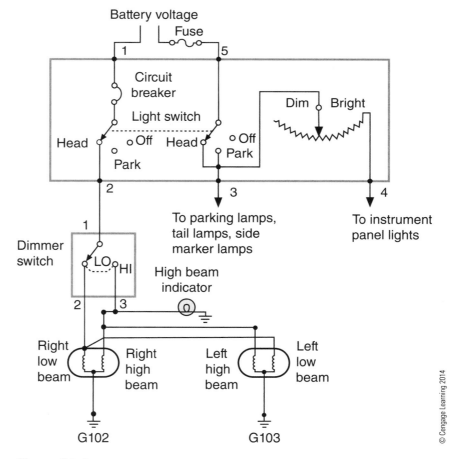

Figure 21-1

27. One headlight being brighter than the other can be caused by a _____ connection at the dim headlight.

28. When replacing the headlight shown in Figure 21-2, Technician A says to loosen and remove screw A. Technician B says to loosen and remove screw B. Who is correct?

    a. Technician A
    b. Technician B
    c. Both A and B
    d. Neither A nor B

Figure 21-2

29. Explain why it is important to not touch the glass bulb when replacing a halogen insert bulb. _____
_____

30. List three methods by which headlights are aimed. _____
_____
_____

31. Explain how to replace a typical parking light bulb. _____
_____

32. All of the rear parking lights are inoperative, but all other rear lights operate correctly. Which is the most likely cause?

    a. Defective brake light switch
    b. Open park light fuse
    c. All bulbs are burned out
    d. High resistance in the ground circuit

33. A vehicle's turn signals flash faster on the right side than on the left side. Which is the LEAST likely cause?

    a. Burned-out turn signal bulb
    b. Defective flasher unit
    c. Incorrect bulb installed in turn signal socket
    d. Resistance in the turn signal circuit

34. The brake light system shown in Figure 21-3 is inoperative. Technician A says to check for power at terminal B of the brake light switch with the brake pedal depressed. Technician B says an open at terminal G of the turn signal switch could be the problem. Who is correct?
   a. Technician A
   b. Technician B
   c. Both A and B
   d. Neither A nor B

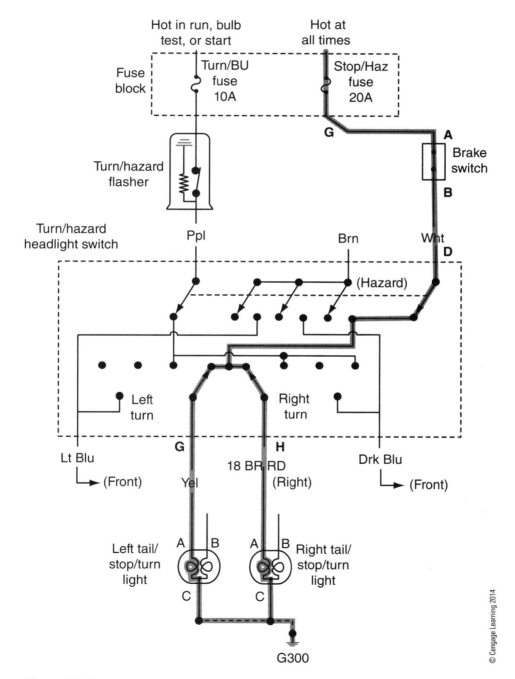

Figure 21-3

35. The _____ panel is part of the dash and contains the gauges, warning lights, and message centers that provide information to the driver.

36. Explain the difference between a yellow dash warning light and a red dash warning light. _____

37. When the ignition is turned to "on" or the start button is pressed, the dash panel performs a _____ _____, which illuminates all of the dash lights momentarily.

38. List nine common warning lights that warn a driver of a fault in a system. _____

39. List three common methods of resetting a maintenance reminder light.
    a. _____
    b. _____
    c. _____

40. The horn circuit in Figure 21-4 is being discussed: The horns do not sound when the horn button is pressed. Technician A says high resistance at splice S105 may be the cause. Technician B says a short to ground at terminal 86 of the horn relay may be the cause. Who is correct?
    a. Technician A
    b. Technician B
    c. Both A and B
    d. Neither A nor B

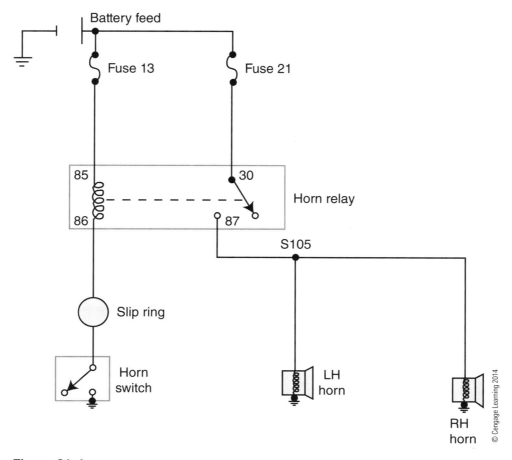

Figure 21-4

41. _____ systems use a remote to activate the power door locks, power sliding doors, and/or rear hatches.

42. Technician A says all remote keyless entry (RKE) systems use a key fob to unlock the doors. Technician B says some RKE systems use a smart key to unlock the doors and start the engine. Who is correct?
    a. Technician A
    b. Technician B
    c. Both A and B
    d. Neither A nor B

43. A vehicle has a complaint of intermittent power window and power door lock operation. Technician A says a faulty master switch may be the cause. Technician B says damaged wiring in the door jamb may be the cause. Who is correct?
    a. Technician A
    b. Technician B
    c. Both A and B
    d. Neither A nor B

44. The left-front window on a vehicle with BCM-controlled power windows does not operate from the door switch but does when controlled with a scan tool. Which is the most likely cause?
    a. Defective power window motor
    b. Defective power window switch
    c. Defective BCM
    d. Open in the power window motor wiring

Activities

1. The high beam lights do not operate in the circuit shown in Figure 21-5. Use the wiring diagram and the color codes shown to examine the circuit and its possible fault.

   Yellow for the component of concern

   Red for constant power to the component

   Orange for switched power to the component

   Green for a variable voltage to the component

   Black for the direct path to ground

   Blue for ground controlled by a switch or other control device

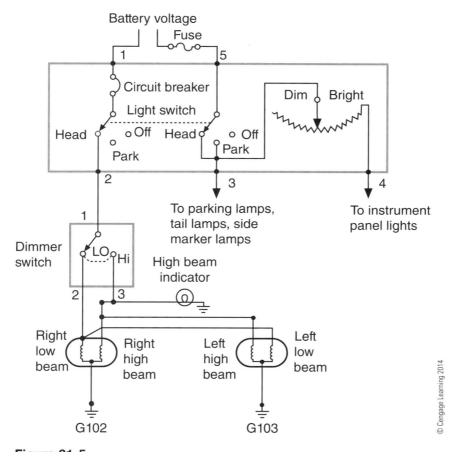

Figure 21-5

INSTRUCTOR VERIFICATION:

2. Obtain a selection of interior and exterior light bulbs from your instructor. Inspect each bulb for signs of damage, broken filaments, and loss of vacuum. Check the resistance of each bulb with a DMM and record your results.

   a. Bulb type _____ Bulb number _____ Resistance _____
      Condition _____
   b. Bulb type _____ Bulb number _____ Resistance _____
      Condition _____
   c. Bulb type _____ Bulb number _____ Resistance _____
      Condition _____
   d. Bulb type _____ Bulb number _____ Resistance _____
      Condition _____
   e. Bulb type _____ Bulb number _____ Resistance _____
      Condition _____

INSTRUCTOR VERIFICATION:

Lab Worksheet 21-1

Name _____ Date _____ Instructor _____

Year _____ Make _____ Model _____

Determine Bulb Applications

1. Remove each type of bulb listed and record the bulb number. Bulb types are often found in the owner's manual or in the service information.
   a. Sealed beam head light _____
   b. Insert head light bulb—low beam _____
   c. Insert head light bulb—high beam _____
   d. Front turn signal/marker light _____
   e. Brake light _____
   f. License plate light _____
   g. Interior dome light _____
   h. Under dash or door panel light _____

2. Why do you think so many different types of bulbs are used in the vehicle? _____
   _____
   _____

3. What problems can result from replacing a bulb with an incorrect bulb type?
   _____
   _____

INSTRUCTOR VERIFICATION:

Lab Worksheet 21-2

Name _____ Date _____ Instructor _____

Year _____ Make _____ Model _____

Using a Scan Tool to Test Lighting System Operation

1. Scan tool used _____

2. Connect and configure the scan tool. Navigate to the system/module for lighting. System/module _____ used.

3. List the exterior lighting systems available. _____
   _____

4. List the interior lighting circuits available. _____
   _____

5. Select several lighting circuits to perform active commands. Turn each circuit on and note its operation.
   Circuit _____ Operation correct _____
   Circuit _____ Operation correct _____
   Circuit _____ Operation correct _____

6. If a bulb failed to operate, what would be your next step in diagnosis and why?
   _____
   _____
   _____

7. If an entire lighting system failed to operate, what would be your next step in diagnosis?
   _____
   _____
   _____

INSTRUCTOR VERIFICATION:

Lab Worksheet 21-3

Name _____ Date _____ Instructor _____

Year _____ Make _____ Model _____

Wiper/washer inspected _____ Front _____ Rear

Inspect Wiper/Washer Operation

1. Perform a visual inspection of the windshield wiper blades. Note your findings.
   _____

2. Turn the ignition on and activate the front windshield washer.

Washer pump operation	_____ OK	_____ Not OK
Washer spray amount	_____ OK	_____ Not OK
Washer spray pattern	_____ OK	_____ Not OK
Wiper effectiveness	_____ OK	_____ Not OK
Wiper noise	_____ OK	_____ Not OK

3. Turn the ignition on and activate the rear windshield washer.

Washer pump operation	_____ OK	_____ Not OK
Washer spray amount	_____ OK	_____ Not OK
Washer spray pattern	_____ OK	_____ Not OK
Wiper effectiveness	_____ OK	_____ Not OK
Wiper noise	_____ OK	_____ Not OK

4. Based on your inspection, what is the condition of the wipers/washer? _____
   _____
   _____

INSTRUCTOR VERIFICATION:

Lab Worksheet 21-4

Name _____ Date _____ Instructor _____

Year _____ Make _____ Model _____

Scan tool used _____

Check Module Status

1. Connect the scan tool and turn the ignition on. Does the scan tool power up correctly?
   Yes _____   No _____

   If no, what does this indicate? _____

2. Does the scan tool support a module or network status test?
   Yes _____   No _____

   If yes, what is the test called? _____

3. If the scan tool supports a module/network test, perform the test and record the results.
   _____
   _____
   _____

4. If the scan tool does not support module/network tests, attempt to communicate with modules known to be present on the vehicle. Record your results. _____
   _____
   _____

5. Based on your testing, are all modules active on the network?
   Yes _____   No _____

6. Do any of the modules have stored DTCs?   Yes _____ No _____

   If yes, record the DTCs. _____

7. Select an accessory system such as lighting for testing. System selected _____

8. List any bidirectional (active) command functions available in the system selected.
   _____
   _____

9. How can this tool capability be used to diagnose and repair a problem with the vehicle? _____
   _____

INSTRUCTOR VERIFICATION:

CHAPTER 22
Engine Performance Principles

Review Questions

1. The dominant types of engines in use today are the gasoline-powered _____ cycle engine and the _____ engine.

2. In a gasoline-powered engine, the air and fuel mixture is ignited by a high-voltage spark delivered to the _____ chamber.

3. A diesel engine _____ air in the cylinder so much that the heat ignites the fuel injected into the cylinder.

4. At their most basic, internal combustion engines are _____ pumps.

5. At sea level, _____ pressure is 14.7 pounds per square inch.

6. Atmospheric pressure _____ as altitude increases.

7. All of the following statements about pressure and vacuum are correct except:
   a. Downward piston movement increases pressure in the cylinder.
   b. The lower pressure in the engine is called vacuum.
   c. The difference between outside and inside the cylinder causes air to flow into the engine.
   d. All are correct.

8. The pressure inside the engine that is lower than atmospheric pressure is called _____.

9. A tire pressure gauge is calibrated to read _____ pressure at normal atmospheric pressure.

10. Explain why atmospheric pressure and vacuum are necessary for an engine to run.
    _____
    _____
    _____
    _____

413

11. Both gasoline and diesel fuel are made of _____.

12. Gasoline contains many _____ to make it usable as a fuel.

13. Diesel fuel can contain certain types of _____, which not only survive in the fuel but also actually feed on it.

14. To be able to extract the chemical energy stored in gasoline, the air-fuel mixture is _____, forcing the molecules closer together.

15. How much the air-fuel charge is compressed depends on the _____ ratio of the engine.

16. Uncontrolled _____ causes reduced power, poor performance, increased emissions, and in some cases, engine damage.

17. The reciprocating motion of the pistons must be converted into _____ motion.

18. List the four cycles of the internal combustion engine in order.
    a. _____
    b. _____
    c. _____
    d. _____

19. All current production gasoline engines for automotive use in the United States are _____ -cycle, _____-cooled, _____ ignition engines.

20. All of the following statements about diesel engines are correct except:
    a. Automotive diesels are liquid-cooled.
    b. Diesel engines do not use spark plugs.
    c. Diesel engines are smaller and lighter than gasoline engines.
    d. Diesel engines use four-cycle operation.

21. In an Atkinson cycle engine, the _____ stroke is extended, allowing some of the air-fuel mixture to move back up into the intake manifold.

22. The type of internal combustion engine that does not use reciprocating parts is called the:
    a. Diesel engine.
    b. Rotary engine.
    c. Miller engine.
    d. All engines use reciprocating parts.

23. All of these construction materials are used in modern engine design to save weight except
    a. Aluminum.
    b. Plastic.
    c. Cast iron.
    d. Magnesium.

24. Identify the components labeled in Figure 22-1.
    a. ____________________ b. ____________________
    c. ____________________ d. ____________________
    e. ____________________ f. ____________________

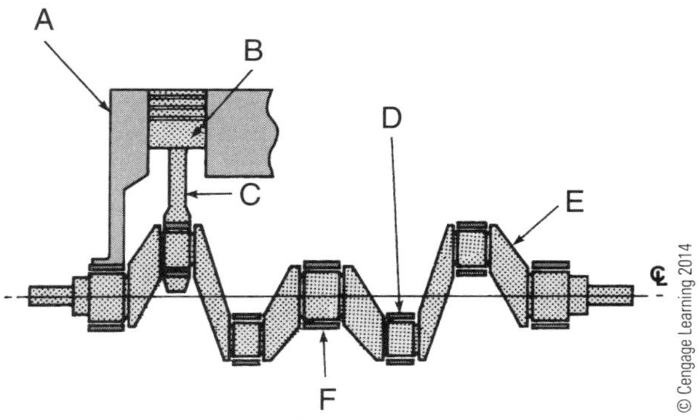

Figure 22-1

25. List six components of the engine bottom end.
    a. ____
    b. ____
    c. ____
    d. ____
    e. ____
    f. ____

26. Which of the following is not a function of the piston?
    a. Compress the air-fuel mixture.
    b. Transfer movement to the crankshaft.
    c. Seal the top of the combustion chamber.
    d. Draw air into the cylinder.

27. In an overhead valve engine, the camshaft is located in the ____________________.

© 2014 Cengage Learning. All Rights Reserved. May not be scanned, copied or duplicated, or posted to a publicly accessible web site, in whole or in part.

416 Chapter 22 Engine Performance Principles

28. Describe the differences between a wet sump and a dry sump oil pump. _____
_____
_____

29. List six components of the engine top end. _____
_____
_____
_____
_____

30. Label the parts shown in Figure 22-2.
    a. _____   b. _____
    c. _____   d. _____
    e. _____   f. _____
    g. _____

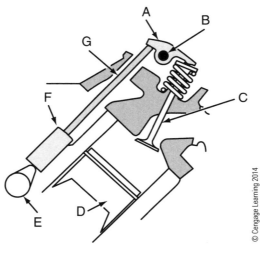

Figure 22-2

31. Explain the operation of the overhead valvetrain. _____
_____
_____
_____

Chapter 22 Engine Performance Principles 417

32. Explain the operation of the overhead cam valvetrain. _____
_____
_____
_____
_____

33. Getting the air into the engine is the responsibility of the _____ system.

34. During low-speed operation, _____ intake runners improve power by speeding up the airflow.

35. A turbocharger is driven by:
    a. Drive belt.
    b. Electric motor.
    c. Exhaust gases.
    d. Intake airflow.

36. Turbocharged and supercharged engines use a(n) _____ to remove some of the heat from the air entering the engine.

37. List four functions of the exhaust system. _____
_____
_____
_____

38. The lubrication system is used to remove _____, clean the inside of the engine, and trap _____.

39. On some diesel engines, engine oil may be used to open and close the _____.

40. Explain three functions of the cooling system. _____
_____
_____
_____

41. Coolant is a mixture of _____-_____ and water.

42. Explain why a pressurized cooling system is used on modern engines.
_____
_____
_____
_____

418 Chapter 22 Engine Performance Principles

43. Label the parts of the cooling system shown in Figure 22-3.

   a. _____   b. _____
   c. _____   d. _____
   e. _____   f. _____
   g. _____

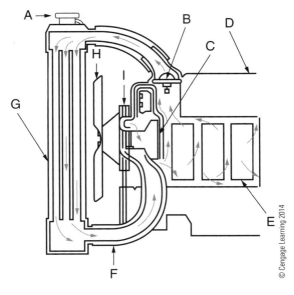

Figure 22-3

44. The _____ system on a gasoline-powered engine provides the heat to ignite the air-fuel mixture in the combustion chamber.

45. All modern gasoline engines use _____ to deliver the gasoline to the cylinders.

46. Describe in detail what happens during the intake stroke of a gasoline engine.

   _____
   _____
   _____
   _____

47. Describe in detail what happens during the compression stroke of a gasoline engine.

   _____
   _____
   _____
   _____

Chapter 22 Engine Performance Principles   419

48. Describe in detail what happens during the combustion stroke of a gasoline engine.

___

49. Describe in detail what happens during the exhaust stroke of a gasoline engine.

___

50. Horsepower is the rate of the amount of _____ performed in a specific amount of time.

51. Gasoline-powered engines are approximately _____ efficient.
    a. 80 percent
    b. 35 percent
    c. 20 percent
    d. 10 percent

52. Explain the fuel efficiency advantage of the Atkinson cycle engine.

___

53. Hybrid-electric vehicles (HEVs) combine ICEs and powerful electric motor/generators to _____ the vehicle, recapture _____ energy, increase fuel economy, and reduce _____ emissions.

Activities

I. Displacement

Engine size is determined by cylinder displacement, which is the total volume of the cylinders and combustion chambers. To understand displacement, examine the volume of a basic cylinder, shown in Figure 22-4. When the piston is at BDC, the cylinder can hold a certain amount of liquid, represented by V. To determine the value of V, we need to know the radius of the cylinder and its depth. Radius is one-half of the diameter, which is the distance across the cylinder from side-to-side. Depth is the total height from the top edge to the bottom of the cylinder, which is also the top of the piston.

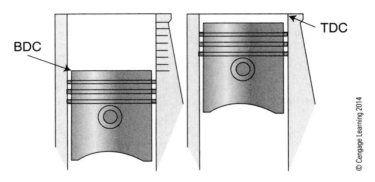

Figure 22-4

If the cylinder is 3.5 in. (88.9 mm) in diameter and 3.5 in. (88.9 mm) deep, this would equal a bore of 3.5 in. and a stroke of 3.5 in. Bore and stroke are the terms used to describe cylinder diameter and depth, as shown in Figure 22-5. To calculate the volume, the bore of 3.5 in. is converted to radius. Since radius is one-half of diameter (or bore), then a bore of 3.5 in. has a radius of 1.75 in. (44.45 mm). Using the formula $V = \pi r^2 h$, where V is the total volume, π is the value of pi, or 3.14159, r^2 is the cylinder radius squared, and h is the height (or stroke) of the cylinder, the volume can be determined. Our equation will look like $V = 3.14159 \cdot 1.75^2 \cdot 3.5$. The cylinder volume, or displacement, equals about 33.67 cubic inches, or 33.7^3 in. In metric measurements, the cylinder displaces about 552 cubic centimeters or 552 cc. If our engine has eight cylinders, then we multiply $8 \cdot 33.7^3$ in. to get a total engine displacement of 269.39 cubic inches or 4416 cc.

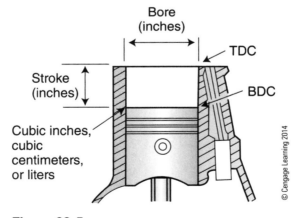

Figure 22-5

Another method of calculating displacement is by using the formula: .785 · Bore · Stroke · number of cylinders. Since .785 is one-quarter of π, some people find it easier to remember this way since the actual bore measurement can be used.

To measure displacement, obtain from your instructor a cylindrical container that you can measure for diameter and depth. Using a telescoping gauge and depth gauge, measure the inside diameter and depth of the container and record. Calculate the volume of the container using the formulas above.

1. Inside diameter _____ Depth _____ Volume _____

Using the volume you found, multiply the total by 4, 6, and 8 to get an idea of what the amount equates to for a four-cylinder, six-cylinder, and eight-cylinder engine.

2. Four-cylinder _____ Six-cylinder _____ Eight-cylinder _____

Compare these numbers with the engine sizes found in modern cars and light trucks.

3. How does the displacement compare to the external size of an engine? _____

4. What are some factors that limit engine displacement? _____

5. Can engine wear affect displacement? Why/how? _____

6. If an engine is bored oversize to correct for cylinder wear, what effect will this have on displacement? _____

7. How would carbon buildup on a piston affect displacement? _____

II. Compression Ratio

Compression ratio refers to the volume of the cylinder when the piston is at BDC compared to the volume when the piston is at TDC, shown in Figure 22-6. The difference in volume is the compression ratio.

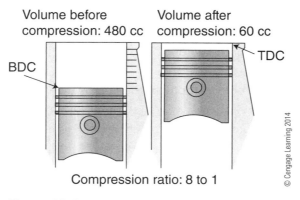

Figure 22-6

Obtain from your instructor a syringe that holds 100 cc of liquid and has graduated markings along the length marking every 10 cc, like that shown in Figure 22-7. Pull the plunger back, filling the syringe with 100 cc of air. This is similar to filling the cylinder of an engine when the piston has moved from TDC to BDC on the intake stroke.

422 Chapter 22 Engine Performance Principles

Now, cap the opening of the syringe and push the plunger back in, compressing the trapped air inside. Notice that as you compress the air, it becomes more difficult to push the plunger. If you can compress the original 100 cc volume of air until the plunger reaches the 10 cc mark, you have compressed the air by a factor of 10, or by a ratio of 10:1. By taking the original volume of 100 cc and reducing it down to 10 cc, you have compressed the air 10 times. Compression ratio = volume 1/volume 2 or 100/10 = 10 or 10:1.

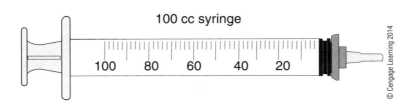

Figure 22-7

1. Why did the plunger become harder to push as the volume of air decreased? _____

2. What would the compression ratio in the syringe be if you compressed the air from 100 cc to 50 cc?
   _____

3. If you compressed the air in the syringe to the 20 cc mark, what would the compression ratio be?
   _____

For a simple experiment, Boyle's Law can be used to determine the pressure in the syringe. P = pressure, P_i = initial pressure, or 14.7 psi, or atmospheric pressure at sea level, V_i is the initial volume, in this case 100 cc, and V_f is the final volume. Use the formula for final pressure or $P_f = P_i V_i / V_f$ and a compression ratio of 2:1 (100 cc to 50 cc):

4. What would be air pressure in the syringe? _____

5. If the air were compressed 10:1, what would the pressure of the air in the syringe be? _____

6. What design factors affect the compression ratio of an engine? _____

7. Why do some engines have higher or lower compression ratios than others? _____

8. Why is the compression ratio of a gasoline-powered engine different from a diesel engine? _____

The actual pressure in an operating cylinder will be different from what you have experienced with this experiment because of the effects of valve timing, intake design, combustion chamber design, and other factors. These experiments do not take into account combustion chamber size, which also has an effect on displacement and compression.

INSTRUCTOR VERIFICATION:

Lab Worksheet 22-1

Name _____ Date _____ Instructor _____

Year _____ Make _____ Model _____

Determine Engine Design and Construction

Determine the following:

1. Engine size   _____ cubic inches   _____ liters

2. Engine type   _____ V-block   _____ Inline   _____ Boxer

                       _____ W-block   _____ Rotary   _____ Other

3. Cylinder bore _____   Stroke _____

4. Compression ratio _____   Firing order _____

5. Horsepower _____   Torque _____ @ _____ rpm

6. Valvetrain type   _____ OHV   _____ OHC   _____ DOHC

7. Fuel type   _____ Gasoline _____ Diesel   _____ Other

8. Fuel delivery system   _____ Port fuel injection   _____ Direct injection

                           _____ Central injection   _____ Other

9. Block construction material _____

10. Cylinder head construction material _____

11. Intake manifold construction material _____

INSTRUCTOR VERIFICATION:

CHAPTER 23
Engine Mechanical Testing and Service

Review Questions

1. Modern engine control systems cannot correct for _____ compression or incorrect _____ timing problems.

2. Before any attempt is made to correct a performance problem, the engine's _____ condition must first be verified.

3. When diagnosing a fluid leak, a _____ light may be used with a special dye to make locating the leak easier.

4. Some vacuum gauges can measure both vacuum and _____ pressure.

5. Scopes display _____ and _____ over _____ time, making them very useful for testing sensors.

6. Explain why you should let the engine cool down before performing services.
   _____
   _____

7. List five different types of fluid leaks that are possible from around the engine compartment.
   a. _____
   b. _____
   c. _____
   d. _____
   e. _____

8. Describe how to distinguish between the different types of fluids you listed in Question 7.
   _____
   _____
   _____
   _____

9. Explain how to perform an engine oil pressure test. _____
   _____
   _____
   _____

10. Technician A says a vacuum leak will affect all cylinders equally. Technician B says a vacuum leak may only affect one cylinder. Who is correct?
    a. Technician A
    b. Technician B
    c. Both A and B
    d. Neither A nor B

11. A vacuum leak will make the air/fuel ratio:
    a. Rich.
    b. Lean.
    c. 14.7:1.
    d. There will be no effect on the air/fuel ratio.

12. A power balance test is performed to:
    a. Determine why a cylinder has low compression.
    b. Determine which cylinder is not producing its share of power.
    c. Locate a vacuum leak.
    d. None of the above.

13. On some vehicles, a _____ tool is used to perform a power balance test.

14. All of the following can cause a compression leak except:
    a. Leaking head gasket.
    b. Burnt exhaust valve.
    c. Broken oil control ring.
    d. Cracked cylinder head.

15. An engine has a low compression reading on one cylinder: Technician A says a wet test should be performed to determine if the problem is with the valves or the rings. Technician B says a wet test will determine if the head gasket is leaking. Who is correct?
    a. Technician A
    b. Technician B
    c. Both A and B
    d. Neither A nor B

16. A _____ compression test is performed using a current clamp and a scope.

17. The waveform shown in Figure 23-1 was captured during a relative compression test on an eight-cylinder engine. Technician A says this indicates normal cranking compression readings. Technician B says a problem is indicated in one cylinder. Who is correct?

    a. Technician A
    b. Technician B
    c. Both A and B
    d. Neither A nor B

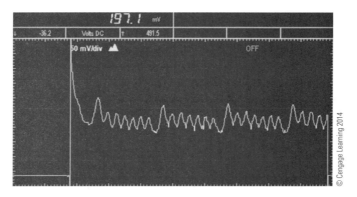

Figure 23-1

18. Explain what conditions can be diagnosed using a running compression test.

    _____
    _____
    _____
    _____

19. A running compression test is often used to pinpoint a problem with:

    a. Worn cam lobes.
    b. Worn piston rings.
    c. Bent valves.
    d. Restricted exhaust.

20. Which of the following types of problems is NOT diagnosed with a cylinder leakage test?

    a. Worn rings
    b. Worn cam lobes
    c. Leaking head gasket
    d. Bent valve

21. Which of the following tests is performed to determine if a cylinder is not producing power?
    a. Compression test
    b. Vacuum test
    c. Power balance test
    d. Cylinder leakage test

22. Bubbles are present in the cooling system during a cylinder leakage test. Describe what can cause this to occur.
    _____
    _____

23. Match the color of exhaust smoke to its cause for a gasoline engine.
    Blue   Combustion/cooling system leak
    White   Excessive fuel
    Black   Excessive oil in cylinder

24. Describe what a typical valvetrain noise sounds like. _____
    _____
    _____

25. An engine has an unusual knocking noise during operation. Which of the following tests should be performed?
    a. Vacuum test
    b. Oil pressure test
    c. Compression test
    d. Power balance test

26. Induction and vacuum leaks usually cause a high-pitched _____ sound.

27. Explain two problems associated with broken powertrain mounts. _____
    _____
    _____
    _____
    _____

28. List four possible causes of engine vibration. _____
    _____
    _____
    _____

29. List three common exhaust system problems. _____

_____

_____

_____

30. Explain how to test for a restricted exhaust system. _____

_____

_____

_____

_____

31. Describe five factors involved in deciding whether to rebuild or replace an engine.

   a. _____
   b. _____
   c. _____
   d. _____
   e. _____

32. Describe the differences between minor and major engine concerns. _____

_____

_____

_____

33. List three reasons why a shop may not perform major engine repairs that require rebuilding an engine.

34. Some engines require periodic valve _____ adjustments.

35. Define valve lash and why it is important that it be checked and adjusted.

_____

_____

_____

_____

36. To measure and set valve lash, a _____ gauge is used between the rocker arm and the valve.

37. Technician A says excessive valve lash may result in a tapping noise with the engine running. Technician B says if the valve lash is too small, the valve may stay open when the engine is cold. Who is correct?
    a. Technician A
    b. Technician B
    c. Both A and B
    d. Neither A nor B

38. Describe what is meant by the term interference engine. _____
_____
_____

39. Describe the general procedures to replace a timing belt. _____
_____
_____
_____
_____

40. _____ are used between a moving and a nonmoving part, and _____ are used between two nonmoving parts.

41. Technician A says steel gasket scrapers are acceptable to use on all types of engine components. Technician B says gasket cleaning discs can safely be used on all types of gasket surfaces. Who is correct?
    a. Technician A
    b. Technician B
    c. Both A and B
    d. Neither A nor B

Lab Worksheet 23-1

Name _____ Date _____ Instructor _____

Year _____ Make _____ Model _____ Engine _____

Cranking Compression Testing—Conventional Compression Tester

1. Disable the fuel and ignition systems. Explain how this is accomplished. _____
   _____

Warning! Hybrid vehicles require special procedures for conducting compression tests. Do not try to perform this test on a hybrid without following the manufacturer's precautions and procedures.

2. Use an air blow gun to clean around the spark plugs. Remove all the spark plugs.

3. Block the throttle open. Explain why the throttle should be wide open when performing a compression test. NOTE: Refer to the service information for procedures if vehicle is equipped with electronic throttle control.
   _____
   _____

4. Install a battery charger and set to a low charge rate. Explain why a battery charger should be used during a compression test. _____
   _____

5. Install the compression gauge into cylinder number 1. Crank the engine for five seconds and note the compression gauge reading. Record the reading in the space below.

Dry test	Wet test	Dry test	Wet test
Cyl No.		Cyl No.	
Cyl No.		Cyl No.	
Cyl No.		Cyl No.	
Cyl No.		Cyl No.	

6. If a cylinder's compression reading is more than 10 percent lower than the others, a wet test should be performed. Squirt a small amount of oil into the low cylinder and repeat the compression test.

7. Based on your testing results, what is the condition of the engine?
   _____
   _____

INSTRUCTOR VERIFICATION:

Lab Worksheet 23-2

Name _____ Date _____ Instructor _____

Year _____ Make _____ Model _____ Engine _____

Running Compression Test—Conventional Compression Tester

1. Record the cylinder firing order: _____

2. Remove the spark plug from the cylinder being tested. Install a compression gauge in the spark plug hole. To monitor actual cylinder pressures, remove the Schrader valve from the compression tester. This allows the needle to rapidly bounce with cylinder pressure changes.

NOTE: Damage to the gauge can result if the needle is pulled against the stop pin during this test. Depending on the type of gauge being used, leave the Schrader valve in the compression tester and use the bleed valve to bleed air from the gauge during the test.

3. Start the engine and allow the reading to stabilize. Needle bounce is normal for a running compression test. Cylinder running pressure _____

4. Snap the throttle wide open and return to idle. Note and record the peak reading. This reading should be higher than the idle reading. Snap throttle pressure _____

5. Record your readings for running and snap compression for all cylinders. The running compression reading should be approximately 50 psi to 75 psi, and snap compression should be about 80 percent of cranking compression.

Running Pressure		Snap Pressure	
1. _____ 5. _____		1. _____ 5. _____	
2. _____ 6. _____		2. _____ 6. _____	
3. _____ 7. _____		3. _____ 7. _____	
4. _____ 8. _____		4. _____ 8. _____	

Worn cam lobes and weak or broken valve springs can cause low running compression. Higher-than normal readings, over 80 percent of cranking compression pressure, can be caused by a restricted exhaust system.

6. Based on your testing, what conclusions can you make about the engine?

_____

_____

INSTRUCTOR VERIFICATION: _____

Chapter 23 Engine Mechanical Testing and Service 435

Lab Worksheet 23-3

Name _____ Date _____ Instructor _____

Year _____ Make _____ Model _____ Engine _____

Relative Compression Testing

1. Disable the fuel and ignition systems. Explain how this is accomplished. _____
   _____

Warning! Hybrid vehicles require special procedures for conducting compression tests. Do not try to perform this test on a hybrid without following the manufacturer's precautions and procedures.

2. Connect a current probe to a lab scope.

3. Set up the current probe and lab scope to display cranking current draw.
   a. Current probe setting _____
   b. Lab scope voltage range _____
   c. Lab scope time base _____

4. Crank the engine and capture the cranking current waveform.

5. Describe or draw the waveform pattern.

6. Determine the peak and lowest amperages during cranking.
   a. Peak amperage _____
   b. Lowest amperage _____

7. Based on your test results, what is the condition of the engine?

INSTRUCTOR VERIFICATION: _____

Lab Worksheet 23-4

Name _____ Date _____ Instructor _____

Year _____ Make _____ Model _____ Engine _____

Compression Testing—Pressure Transducer

1. Connect and calibrate the pressure transducer to the lab scope.
    a. Pressure range selected _____
    b. Time base selected _____

2. Remove the spark plug from the cylinder being tested. Install the pressure transducer in the spark plug hole.

3. Crank the engine and obtain a compression waveform.

    Maximum pressure recorded _____

4. Start the engine and obtain several cycles of cylinder compression.
    a. Maximum cylinder running pressure _____
    b. Minimum cylinder running pressure _____

5. Snap the throttle wide open and return to idle. Note and record the peak reading.

    Snap throttle pressure _____

6. Based on your testing, what is the condition of the cylinder?
    _____
    _____

INSTRUCTOR VERIFICATION:

Chapter 23 Engine Mechanical Testing and Service 439

Lab Worksheet 23-5

Name _____ Date _____ Instructor _____

Year _____ Make _____ Model _____ Engine _____

Cylinder Power Balance Test with a Scan Tool

1. Obtain a scan tool applicable for the vehicle being tested.

    Scan tool used _____

2. Connect the scan tool to the DLC and enter into powertrain diagnostics. Locate the power balance test.

3. Initial engine rpm _____

4. Follow the procedure indicated by the scan tool, disable each cylinder, and note the rpm drop:

    Cyl 1 _____   Cyl 5 _____

    Cyl 2 _____   Cyl 6 _____

    Cyl 3 _____   Cyl 7 _____

    Cyl 4 _____   Cyl 8 _____

5. What is the rpm difference between the weakest and strongest cylinders? _____

6. Based on the test results, what is the necessary action? _____

INSTRUCTOR VERIFICATION:

Lab Worksheet 23-6

Name _____ Date _____ Instructor _____

Year _____ Make _____ Model _____ Engine _____

Engine Vacuum Testing

1. Connect a vacuum gauge to a suitable vacuum port on the engine. Describe the location of the port.
   _____

2. With the ignition and/or fuel system disabled, crank the engine and record the vacuum reading.

   Cranking vacuum reading _____

   Reconnect the ignition and/or fuel system.

3. Start the engine and allow it to idle. What is the idle vacuum reading? _____

   a. Is the gauge showing a steady needle?   Yes _____   No _____

   b. What would a bouncing or fluctuating needle indicate? _____

4. Quickly snap the throttle wide open and then close. Note the readings at WOT and on deceleration

   WOT _____   Decel _____

5. Increase the engine speed to 2,000 rpm and hold. Note the vacuum reading. _____

6. Does the vacuum reading decrease after one minute?   Yes _____   No _____

7. If the vacuum reading dropped at 2,000 rpm, what would that indicate? _____

8. Based on your testing, what is the condition of the engine?

INSTRUCTOR VERIFICATION:

Lab Worksheet 23-7

Name _____ Date _____ Instructor _____

Year _____ Make _____ Model _____ Engine _____

Cylinder Leakage Testing

1. With the engine cool, remove the spark plug from the cylinder to be tested.

2. Connect the cylinder leak tester to the shop air and zero out the tester. If this step is not performed, the test results will not be accurate.

    Instructor's check _____

3. Make sure the Schrader valve is removed from the cylinder hose and install the hose into the spark plug hole.

4. Set the engine so that the cylinder being tested is at TDC compression. You may need to prevent the crankshaft from spinning by using a socket and ratchet to hold the crank pulley bolt.

    Instructor's check _____

5. Connect the leakage tester hose to the cylinder hose and note the reading on the gauge.

    Gauge reading _____

6. Is the reading excessive? _____ Yes _____ No

    If yes, try to determine where the air is escaping from. Note your results.
    _____

7. Based on your testing, what is the condition of the cylinder? _____
    _____

INSTRUCTOR VERIFICATION:

CHAPTER 24

Engine Performance Service

Review Questions

1. Until the 1980s, engine and transmission systems were, in nearly all cars and trucks, nearly 100 percent _____ operated.

2. Even with on-board computers to manage engine operation, some items, such as _____ filters, spark plugs, and other normal wear items, still require periodic _____ and replacement.

3. The _____ of gasoline's vapors makes working around gasoline hazardous.

4. List five safety precautions for working with gasoline.
   a. _____
   b. _____
   c. _____
   d. _____
   e. _____

5. _____ fuel filters have either plastic or steel shells and contain a _____ filter element.

6. Technician A says all modern vehicles have a serviceable inline fuel filter. Technician B says some vehicles place the fuel filter in the fuel tank and it is routinely replaced as part of a maintenance program. Who is correct?
   a. Technician A
   b. Technician B
   c. Both A and B
   d. Neither A nor B

7. Before you begin to replace a fuel filter, the fuel system must be _____ to prevent fuel from spraying from an open line.

445

8. Technician A says special tools may be required to replace a fuel filter. Technician B says fuel filters are directional and must be installed a certain way. Who is correct?
   a. Technician A
   b. Technician B
   c. Both A and B
   d. Neither A nor B

9. Identify the parts of the spark plug shown in Figure 24-1.
   a. ____________________ b. ____________________
   c. ____________________ d. ____________________
   e. ____________________ f. ____________________
   g. ____________________

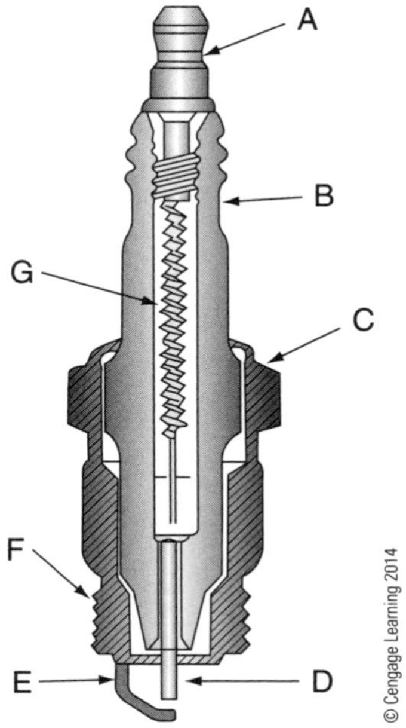

Figure 24-1

10. Which is NOT a typical spark plug replacement interval?
    a. 30,000 miles
    b. 60,000 miles
    c. 100,000 miles
    d. All of the above

11. Describe why it is important to let the engine cool before removing the spark plugs.

____

____

____

© 2014 Cengage Learning. All Rights Reserved. May not be scanned, copied or duplicated, or posted to a publicly accessible web site, in whole or in part.

12. Explain what you should do if a spark plug is very difficult to remove. _____
_____
_____
_____

13. List three things spark plugs should be inspected for. _____
_____
_____

14. Technician A says antiseize should be used on all new spark plugs. Technician B says all spark plugs should be torqued to specifications. Who is correct?
    a. Technician A
    b. Technician B
    c. Both A and B
    d. Neither A nor B

15. Spark plug gap specifications are typically between _____ and _____.

16. Caution must be used when you are checking and gapping plugs with _____ and _____ electrodes as these are easily damaged.

17. Explain how to properly reinstall and tighten a spark plug. _____
_____
_____
_____

18. Explain how the PCV valve is used to reduce emissions. _____
_____
_____
_____

19. If left unchecked, the crankcase vapors will build up _____, causing _____ leaks.

20. Technician A says at idle, the PCV valve is fully open and flowing vapors back into the intake manifold. Technician B says the PCV valve is closed under low vacuum conditions. Who is correct?
    a. Technician A
    b. Technician B
    c. Both A and B
    d. Neither A nor B

Chapter 24 Engine Performance Service

21. Describe how to test a PCV valve. _____
_____
_____
_____

22. In addition to checking the PCV valve and hose, check the _____ air hose for cracks, rotting, and tight connections.

23. OBD I systems were used until _____.

24. OBD I arose from the need to reduce exhaust _____.

25. Describe two significant differences between OBD I and OBD II systems.
    a. _____
    _____
    b. _____
    _____

26. List three items standardized by OBD II.
    a. _____
    b. _____
    c. _____

27. A major requirement of OBD II is to illuminate the malfunction indicator light if emissions increase _____ percent over the federal test parameter for that vehicle.

28. List eight systems monitored by OBD II.
    a. _____
    b. _____
    c. _____
    d. _____
    e. _____
    f. _____
    g. _____
    h. _____

29. Describe the differences between global and enhanced data. _____
_____
_____
_____

30. A(n) _____ is a test run by the ECM on components and systems to test their operation and to determine that certain operating conditions have been met.

31. Explain how passive, active, and intrusive tests differ from each other. _____
_____
_____
_____

32. Technician A says the CCM is used to monitor sensor information and determine if wiring faults are present. Technician B says the CCM is used to actively test component operation. Who is correct?

    a. Technician A
    b. Technician B
    c. Both A and B
    d. Neither A nor B

33. Which component is used to determine engine misfire?

    a. Throttle position sensor
    b. Crankshaft position sensor
    c. Engine coolant temperature sensor
    d. All of the above

34. A misfire code P0304 indicates a misfire on cylinder number _____.

35. Type A misfires must be detected within _____ to _____ crankshaft revolutions.

36. Explain what is meant by closed-loop operation. _____
_____
_____
_____

37. Fuel control is represented by _____ term and _____ term fuel trim on the scan tool.

38. Technician A says an STFT and LTFT reading of 18 percent indicates the engine is running lean and the computer is adding fuel to the mixture. Technician B says a vacuum leak may cause the lean condition that the computer is adding fuel to compensate for. Who is correct?

    a. Technician A
    b. Technician B
    c. Both A and B
    d. Neither A nor B

39. Explain why some monitors, such as EVAP and catalyst efficiency, are noncontinuous monitors. _____
_____
_____

40. List three oxygen sensor functions the oxygen sensor monitor tests for.

   a. _____
   b. _____
   c. _____

41. Technician A says the catalyst efficiency monitor is dependent upon the oxygen sensor operation. Technician B says a faulty oxygen sensor may cause the catalyst monitor not to complete. Who is correct?

   a. Technician A
   b. Technician B
   c. Both A and B
   d. Neither A nor B

42. Describe the purpose of the EVAP system. _____
_____
_____
_____

43. The EGR system is used to reduce _____ _____ temperatures.

44. A drive cycle is a series of trips designed to allow all _____ to run.

45. Explain enable criteria as related to a trip. _____
_____
_____
_____

46. Regarding the OBD II diagnostic trouble code P0301: Technician A says a P code indicates a powertrain code. Technician B says that P0 indicates the code is generic. Who is correct?

   a. Technician A
   b. Technician B
   c. Both A and B
   d. Neither A nor B

47. A _____ code appears when the ECM has detected a problem but is waiting for a second trip to confirm the code.

48. Describe generic OBD II Mode 10 permanent trouble codes. _____
   _____
   _____
   _____

49. If the VIN information is incorrectly entered into a scan tool during setup, which of the following may result?
   a. No communication
   b. Unusual data
   c. Damage to the scan tool
   d. Both a and b

50. It is likely that OBD III will incorporate some form of _____ monitoring of the vehicle's emission systems.

Chapter 24 Engine Performance Service 453

Lab Worksheet 24-1

Name _____ Date _____ Instructor _____

Year _____ Make _____ Model _____

Engine _____ Fuel system type _____

Relieve Fuel System Pressure

1. Locate and record the manufacturer's procedures for relieving fuel pressure. _____
   _____
   _____

   If the manufacturer does not specify a method to relieve fuel pressure, select one of the following as the correct procedure:

   a. Install a fuel pressure gauge on the pressure test port and release the fuel from the gauge into a suitable container.
   b. Remove the fuel pump fuse, start the engine, and idle until it stalls.
   c. Remove the fuel pump relay, start the engine, and idle until it stalls.
   d. Disconnect the fuel pump connection, start the engine, and idle until it stalls.

2. Method selected _____

   Before any work is performed on the fuel system, any remaining fuel pressure must be relieved.

3. Once work is complete, restore the fuel system to its normal condition.

4. Turn the key on and check for fuel leaks before starting the engine.

5. Start the engine and verify proper fuel system operation.

   Instructor's check _____

INSTRUCTOR VERIFICATION:

© 2014 Cengage Learning. All Rights Reserved. May not be scanned, copied or duplicated, or posted to a publicly accessible web site, in whole or in part.

Lab Worksheet 24-2

Name _____ Date _____ Instructor _____

Year _____ Make _____ Model _____

Engine _____ Number of cylinders _____

Remove, Inspect, and Replace Spark Plugs

1. Locate and record the manufacturer's service procedures for removing and installing spark plugs.
   _____
   _____
   _____
   _____

2. Locate the engine's firing order and note it here. _____

3. Locate the engine's cylinder arrangement and draw it here.

4. Before beginning, make sure the engine is cool to the touch. Do not attempt to remove spark plugs from a hot engine.

5. Carefully remove the spark plug wire or coil from the spark plug. Note any difficulties.
   _____
   _____

Chapter 24 Engine Performance Service

6. Note the size of the spark plug socket needed to remove the spark plug. _____

7. Install the spark plug socket onto the spark plug. Using a ratchet, break the plug loose and remove it from the engine. Note any difficulties. _____

8. Examine the spark plug. Note the condition of the following:
   a. Plug wire terminal _____
   b. Insulation _____
   c. Steel shell and hex _____
   d. Threads _____
   e. Center and ground electrodes _____
   f. Color of the electrodes and insulator _____

9. Using a spark plug gapping tool, carefully measure and record the plug gap.

   Measured gap _____   Gap specification _____

10. Determine if antiseize is used on the spark plug thread.   _____ Yes   _____ No

11. Locate and record the spark plug torque specification. _____

12. Carefully thread the spark plug into the cylinder head. Hand-tighten the plug until it seats against the head and then torque the plug to specifications.

    Instructor's check _____

13. Reinstall the spark plug wire or coil.

    Instructor's check _____

INSTRUCTOR VERIFICATION:

Lab Worksheet 24-3

Name _____ Date _____ Instructor _____

Year _____ Make _____ Model _____

Engine _____

Inspect the PCV System

1. Using the service information, locate and record the manufacturer's procedures for inspecting the PCV valve and system. _____
   _____
   _____
   _____

2. Find and note the location of the PCV valve. _____
   _____

3. With the engine running, carefully remove the PCV valve. Place your thumb over the valve's opening and note the reaction of the valve. _____

4. Reinstall the valve into the engine and inspect the vacuum hose to the valve. Note the condition of the vacuum hose. _____

5. Locate the fresh air hose to the engine for the PCV system. Inspect the air hose and note its condition.
   _____

6. Based on your inspection, what is the condition of the PCV system? _____
   _____
   _____

INSTRUCTOR VERIFICATION:

Lab Worksheet 24-4

Name: Shane Ullgerander Date: 2016/03/01 Instructor: Patrick Rice

Year: ~~2005~~ 2008 Make: ~~Hyundai~~ Chrysler Model: ~~Azera~~ Town & Country

Engine: 6 cyl. 4.0L Scan tool: X-431 PAD II

On-Board Computer Communication Checks

1. Connect a scan tool to the DLC. Select Global OBD II and begin communication. List the communication protocol displayed on the scan tool. **Manufacturer Specific**

2. List the number of ECUs the scan tools detects. **29**

3. Check and record any stored DTCs. **None**

4. Exit Global OBD II and enter enhanced OBD (EOBD) mode. List the systems available on the scan tool menu. (Examples: powertrain, ABS, airbag, etc.) ~~above~~

5. Does the scan tool enable you to perform a network test? Yes ✓ No ___

   If yes, perform the network test and note the results. **Found several faults: SRS system, Powertrain, ABS, Body Control Module**

6. If the scan tool does not have a network test, try to communicate with each system shown from the main menu. Note your results. ___

7. Based on your inspection, is the network operating correctly? Yes ✓ No ~~~~

INSTRUCTOR VERIFICATION:

Chapter 24 Engine Performance Service 461

Lab Worksheet 24-5

Name _____ Date _____ Instructor _____

Year _____ Make _____ Model _____

OBD II Monitor Status

1. Connect a scan tool to the vehicle's DLC. Scan tool used _____

2. Enter Global OBD II and begin communication.

3. Navigate to the OBD monitor status and record the condition of each of the following:

   Misfire _____   Fuel control _____

   Component (CCM) _____ Catalyst _____

   HO_2S _____ O_2 Heater _____

   EVAP _____   EGR _____

   Secondary air _____   PCV _____

   Thermostat _____   Air conditioning _____

4. Are any monitors incomplete or failed?   Yes _____   No _____

   If incomplete, what may be the cause? _____
   _____

   If failed, what may be the cause? _____
   _____

5. Based on your inspection, what conclusion can you make about the OBD system?
   _____
   _____

INSTRUCTOR VERIFICATION: _____

Lab Worksheet 24-6

Name _____ Date _____ Instructor _____

Year _____ Make _____ Model _____

Retrieve DTCs and Freeze Frame Data

1. Connect a scan tool to the vehicle's DLC. Scan tool used _____

2. Navigate to the DTC menu and record any stored DTCs.

    Current DTCs _____

    Pending DTCs _____

    History DTCs _____

3. Select a DTC from the freeze frame data. DTC selected _____

4. Record the following information from the freeze frame record:

    Mileage or starts since first/last fail _____

    Pass/fail counter _____

    Rpm at time of fault _____

    Fuel trims at time of fault _____

    Coolant temperature at time of fault _____

    MAP/MAF at time of fault _____

5. Based on the freeze frame data, what can you determine to be the possible cause of the DTC? _____
_____

INSTRUCTOR VERIFICATION:

CHAPTER 25

Automatic and Manual Transmissions

Review Questions

1. The _____ allows the engine to run at its most efficient speeds so that power is not wasted.

2. List three types of transmissions used in modern vehicles.
   a. _____
   b. _____
   c. _____

3. As on-board computer systems became standard, the _____ transmission also began to be controlled by a powertrain control module or PCM.

4. List three examples of inputs to the PCM used for transmission control.
   a. _____
   b. _____
   c. _____

5. What are three benefits of having the computer control transmission operation?
   a. _____
   b. _____
   c. _____

6. List six major components of a modern automatic transmission.
   a. _____
   b. _____
   c. _____
   d. _____
   e. _____
   f. _____

465

© 2014 Cengage Learning. All Rights Reserved. May not be scanned, copied or duplicated, or posted to a publicly accessible web site, in whole or in part.

Chapter 25 Automatic and Manual Transmissions

7. The _____ converter is used to join the engine's crankshaft to the input shaft of an automatic transmission.

8. Label the components shown in Figure 25-1.
   a. _____
   b. _____
   c. _____

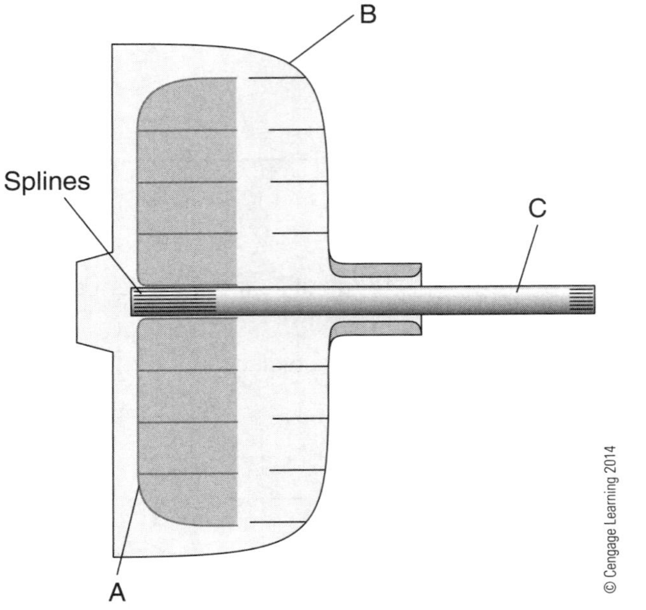

Figure 25-1

9. The torque converter uses _____ to transmit power.

10. Which torque converter component is used to drive the input shaft?
    a. Impeller
    b. Stator
    c. Turbine
    d. Shell

11. Which device is used to achieve optimal torque converter efficiency at cruising speed?
    a. Stator
    b. Lock-up clutch
    c. Oil pump
    d. None of the above

12. The torque converter is only about _____ efficient until the lock-up clutch applies.

13. The rear of the torque converter drives the front transmission _____ pump.

14. Label the parts of the planetary gearset shown in Figure 25-2.

   a. _____ b. _____
   c. _____ d. _____
   e. _____

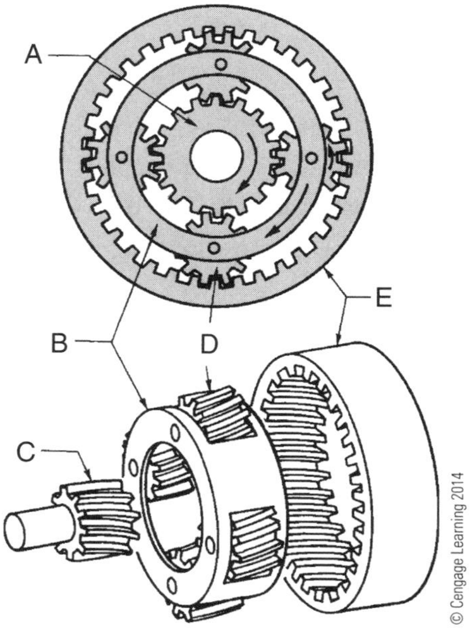

Figure 25-2

15. Describe the operation of a clutch pack in an automatic transmission. _____
_____
_____

16. Before the use of electrically operated transmission shift _____ became standard, all of the shifting functions were controlled by the _____ system.

17. A _____ solenoid is used to open and close a hydraulic passage.

18. Describe how a CVT transmits power from the engine to the drive wheels. _____
_____
_____

Chapter 25 Automatic and Manual Transmissions

19. Technician A says CVT transmissions are used to maintain efficient engine speeds. Technician B says some CVT vehicles are programed to have shift points and manual shifting modes. Who is correct?

    a. Technician A
    b. Technician B
    c. Both A and B
    d. Neither A nor B

20. Some hybrid vehicles use a _____ _____ device instead of a conventional transmission.

21. With a manual transmission, a _____ connects the engine to the transmission and transmits the power from the crankshaft to the transmission.

22. Label the components of the clutch system shown in Figure 25-3.

    a. _____   b. _____
    c. _____   d. _____
    e. _____   f. _____
    g. _____

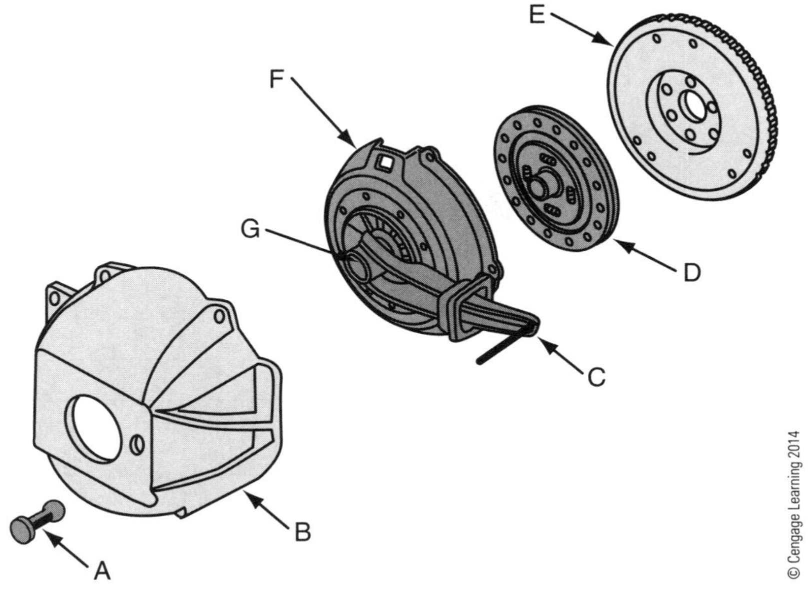

Figure 25-3

23. Explain the purpose and operation of the pressure plate. _____
_____
_____
_____
_____

24. Describe two types of clutch release bearings. _____
_____
_____
_____
_____

25. _____ gear ratios increase torque and allow for quick acceleration.

26. _____ gear ratios allow lower engine speed and economy but with low torque output.

27. List the five major components of a manual transmission.
    a. _____
    b. _____
    c. _____
    d. _____
    e. _____

28. Define gear ratio. _____
_____
_____

29. Match the following gear ratios with the gear number.
    .8:1   Second gear
    1:1   Third gear
    1.3:1   Overdrive (fifth gear)
    2:1   First gear
    3.5:1   Fourth gear

30. Explain the purpose and operation of a synchronizer. _____
_____
_____

31. The shift linkage connects the gear shifter to the _____ shift valve.

32. List five common transmission services.
    a. _____
    b. _____
    c. _____
    d. _____
    e. _____

Chapter 25 Automatic and Manual Transmissions

33. To accurately check the transmission fluid on a vehicle with a dipstick, the fluid usually must be at _____ temperature.

34. Technician A says most modern automatic transmissions use the same type of transmission fluid. Technician B says most automatic transmissions have specific fluid requirements. Who is correct?
    a. Technician A
    b. Technician B
    c. Both A and B
    d. Neither A nor B

35. Technician A says some automatic transmissions have a drain plug and do not have a filter that requires periodic replacement. Technician B says some automatic transmissions do not have a dipstick for checking fluid level. Who is correct?
    a. Technician A
    b. Technician B
    c. Both A and B
    d. Neither A nor B

36. List four problems that can be caused by a faulty transmission range switch.
    a. _____
    b. _____
    c. _____
    d. _____

37. List four problems that can be caused by worn transmission mounts.
    a. _____
    b. _____
    c. _____
    d. _____

38. Describe how to check for worn powertrain mounts. _____

39. When a leak occurs from a manual transmission, check for a plugged _____, which can allow pressure to increase inside the transmission.

40. Explain how to bleed a hydraulic clutch system. _____

41. _____ _____ differentials generally require a specific type of lubricant be used, which contain an additive of special friction modifiers.

42. When an outer CV joint is worn excessively, it will cause a loud _____ sound during turns.

43. Describe how to replace a FWD axle shaft. _____

44. An older 4WD vehicle may have _____ locking hubs.

45. What is indicated by a clicking or ratcheting sound from a 4WD front hub?

Activities

A gear is a circular component that transmits rotational force to another gear or component. Gears have teeth so that they can mesh with other gears without slipping. In Figure 25-4, the driving or input gear has a radius of 1 foot and the driven or output gear has a radius of 2 feet. In this example, the driving gear is turning with 10 pounds of force, which results in 20 pounds of force provided by the driven gear. The output force is increased due to the larger size of the driven gear as each tooth acts as a lever. Since the driven gear has twice as many teeth as the driving gear, the driven gear will turn at one-half of the speed of the driving gear. This equals a gear ratio of 2:1, with two turns of the driving gear to one turn of the driven gear.

Determine the gear ratios for the gear arrangements shown in Figure 25-5.

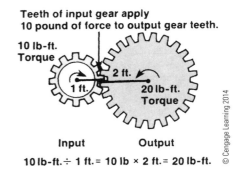

Figure 25-4

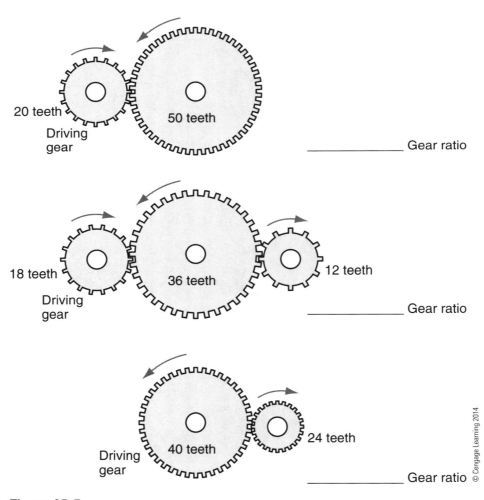

_____ Gear ratio

_____ Gear ratio

_____ Gear ratio

Figure 25-5

INSTRUCTOR VERIFICATION: _____

Lab Worksheet 25-1

Name _____ Date _____ Instructor _____

Year _____ Make _____ Model _____

Engine _____

Identify Automatic Transmission

1. Raise and support the vehicle.

2. Locate the transmission and look for any identification tags.

   Is there an ID tag?   Yes _____   No _____

   a. If yes, what information is contained on the tag? _____
   _____

   b. If no, are there any markings that indicate the transmission manufacturer or other

   information?   Yes _____   No _____

   Manufacturer _____

3. If no ID tag is found, make a drawing of the transmission pan and note the location of the pan bolts.

4. Using the drawing of the transmission pan, identify the pan and transmission using the service information for the vehicle. Transmission model _____

INSTRUCTOR VERIFICATION:

Lab Worksheet 25-2

Name _____ Date _____ Instructor _____

Year _____ Make _____ Model _____

Engine _____ Transmission _____

Checking Automatic Transmission Fluid Type, Condition, and Level

1. Refer to the owner's manual or service information for the recommended fluid type and record it here.
   _____

2. Does the vehicle have a transmission dipstick?   Yes _____   No _____

3. Locate and record the manufacturer's method for checking the fluid level.
   _____
   _____

   a. Does the transmission fluid need to be within a certain temperature range?
      Yes _____   No _____
   b. If yes, operate the vehicle until the fluid is within the specified temperature range.
   c. How will you determine the temperature of the fluid? _____
      _____

4. For vehicles with a dipstick: Remove the dipstick and draw a picture of the end of the stick where the fluid level is checked. Include instructions about how to check the fluid and what type of fluid should be used.

   a. Reinstall and again remove the dipstick; note the fluid level as indicated.
      _____
   b. Describe the appearance of the fluid. _____
      _____

INSTRUCTOR VERIFICATION:

Chapter 25 Automatic and Manual Transmissions

5. For vehicles without a dipstick, determine fluid level and note your results.
   _____

6. Based on your inspection, summarize the condition of the transmission fluid.
   _____
   _____

INSTRUCTOR VERIFICATION:

Lab Worksheet 25-3

Name _____ Date _____ Instructor _____

Year _____ Make _____ Model _____

Engine _____ Transmission _____

Inspecting a Manual Transmission and Checking Fluid Type and Level.

1. Refer to the owner's manual or service information for the recommended fluid type and record it here.
   _____

2. Locate and record the manufacturer's method for checking the fluid level.
   _____
   _____

3. Remove the fluid fill plug and note the level of fluid in the transmission.

   Full _____   Low _____

   a. How did you determine the fluid level? _____
   _____

4. Check the condition of the fill plug and gasket; note any concerns. _____
   _____

5. Reinstall the fill plug and torque it to specifications. Torque spec. _____

6. Visually inspect the transmission for evidence of fluid loss. Note your findings.
   _____

INSTRUCTOR VERIFICATION: _____

Lab Worksheet 25-4

Name _____ Date _____ Instructor _____

Year _____ Make _____ Model _____

Engine _____ RWD _____ 4WD _____

Inspecting a Differential and Checking Fluid Type and Level

1. Refer to the owner's manual or service information for the recommended fluid type and record it here. If inspecting a 4WD vehicle with two differentials, note the fluid type for each.
   _____

2. Locate and record the manufacturer's method for checking the fluid level.
   _____
   _____

3. Remove the differential plug and note the fluid level.

   Full _____   Low _____

   How did you determine the fluid level? _____
   _____

4. Reinstall the fill plug and torque it to specifications (if applicable).

   Torque spec. _____

5. Visually inspect the differential for evidence of fluid loss. Note your findings.
   _____

INSTRUCTOR VERIFICATION:

CHAPTER 26

Heating and Air Conditioning

Review Questions

1. The HVAC system is responsible for maintaining engine _____ and for passenger _____.

2. Describe the components that make up the HVAC system. _____

3. Explain the main functions of the cooling system. _____

4. List eight components of the cooling system.
   a. _____
   b. _____
   c. _____
   d. _____
   e. _____
   f. _____
   g. _____
   h. _____

5. During combustion, temperatures inside the engine can reach _____ °F.

6. _____ is the medium by which heat transfer takes place between the engine and the surrounding air.

Chapter 26 Heating and Air Conditioning

7. Technician A says coolant color is the best method of determining which coolant is used in a vehicle. Technician B says all ethylene-glycol based coolants are interchangeable and can be mixed. Who is correct?
    a. Technician A
    b. Technician B
    c. Both A and B
    d. Neither A nor B

8. Hybrid electric vehicles often use _____ different coolants.

9. The water pump may be driven by the _____ belt or by a(n) _____ drive belt.

10. The _____ transfers the heat of the coolant to the outside air.

11. Explain why pressure is allowed to build up in the cooling system during operation. _____
_____
_____

12. Some cooling systems use a _____ tank, also called a _____ bottle as the fill point for the system.

13. The radiator fan may be driven by a belt, an _____ motor, or hydraulically from the _____ _____ system.

14. Describe the purpose of the fan clutch. _____
_____
_____
_____

15. Explain the purpose of the fan shroud. _____
_____
_____
_____

16. The thermostat uses a _____ pellet to move a piston.

17. Typically thermostats are fully open by _____ °F.

18. A mini radiator, called the _____ _____, is used to supply heat to the passenger compartment.

Chapter 26 Heating and Air Conditioning

19. The _____ system works with both the heating and air condition systems to route air flow for the passenger compartment.

20. List three components found in the HVAC case.
    a. _____
    b. _____
    c. _____

21. _____ air filters are often mounted near the heater core/evaporator housing and are used to trap and prevent _____, _____, and other contaminants from entering the passenger compartment.

22. Explain how stepped resistors are used to control blower motor speeds. _____
_____
_____
_____

23. List three methods of controlling the operation of the blend door.
    a. _____
    b. _____
    c. _____

24. List six components of the air conditioning system.
    a. _____
    b. _____
    c. _____
    d. _____
    e. _____
    f. _____

25. Technician A says refrigerant has a very high boiling point, so it can absorb a lot of heat. Technician B says refrigerant is kept under pressure to prevent it from boiling. Who is correct?
    a. Technician A
    b. Technician B
    c. Both A and B
    d. Neither A nor B

26. The _____ separates the high- and low-pressure sides of the AC system.

27. On hybrid vehicles, the compressor may be driven by a _____ or by high-voltage _____.

484 Chapter 26 Heating and Air Conditioning

28. The high-voltage cables and components used in hybrid vehicles are bright _____ in color to easily distinguish them from other components.

29. An accumulator is used on the _____ pressure side of the system.

30. The component that is used to remove heat from the refrigerant is the _____.

31. To reduce the pressure of the refrigerant entering the evaporator, either an _____ tube or _____ valve is used.

32. Explain four precautions for working with the AC system.
    a. _____
    b. _____
    c. _____
    d. _____

33. Before performing any work on the cooling system, you first must determine what _____ of coolant is required.

34. Describe how to perform a cooling system pressure test. _____
    _____
    _____
    _____
    _____

35. Technician A says a faulty cooling system pressure cap can cause the engine to overheat. Technician B says all cooling system caps should hold at least 15 psi of pressure. Who is correct?
    a. Technician A
    b. Technician B
    c. Both A and B
    d. Neither A nor B

36. Describe how to test for coolant leaking into the combustion chamber. _____
    _____
    _____
    _____
    _____

37. Explain how to inspect cooling system hoses and hose clamps. _____
_____
_____
_____
_____

38. Technician A says a defective thermostat can cause the engine to overheat. Technician B says a defective thermostat can cause the engine to run too cold. Who is correct?
    a. Technician A
    b. Technician B
    c. Both A and B
    d. Neither A nor B

39. A faulty thermostat can cause the _____ _____ light to come on and cause a diagnostic _____ _____ to be set in the computer.

40. _____ the cooling system removes the old coolant and circulates clean water through the system to remove rust and corrosion.

41. To remove trapped _____ from the cooling system, many systems are equipped with a bleed valve.

42. Describe what problems the AC drive belt should be inspected for. _____
_____
_____
_____
_____

43. Cabin air filters may be located near the _____ air inlet in the cowl or in the dash, near the blower motor.

44. During AC operation, _____ forms on the evaporator, which can cause mildew and odor from the HVAC system.

486 Chapter 26 Heating and Air Conditioning

Activities

1. Label the components of the cooling system shown in Figure 26-1.

   a. _____ b. _____
   c. _____ d. _____
   e. _____ f. _____
   g. _____

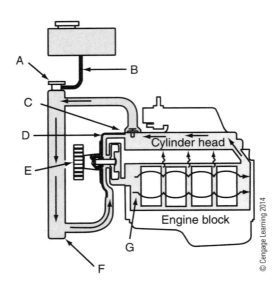

Figure 26-1

2. Label the high and low sections of the AC system shown in Figure 26-2.

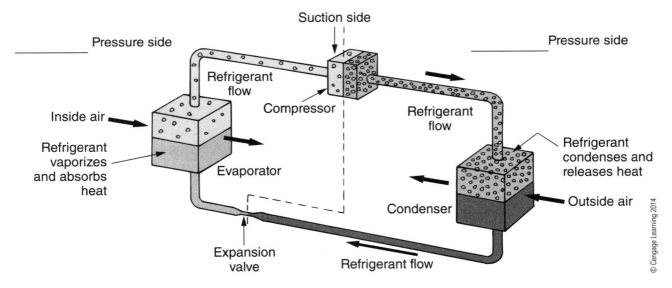

Figure 26-2

INSTRUCTOR VERIFICATION: _____

3. Label the components of the AC system shown in Figure 26-3.

   a. _____   b. _____

   c. _____   d. _____

   e. _____

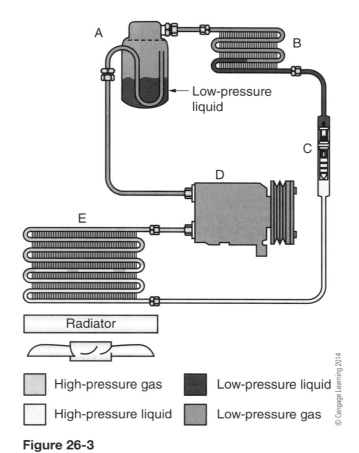

Figure 26-3

INSTRUCTOR VERIFICATION:

Lab Worksheet 26-1

Name _____ Date _____ Instructor _____

Year _____ Make _____ Model _____

Perform a Visual Inspection of the Cooling System

Inspect the following components and note your findings.

1. Pressure cap rating _____ psi   Bar (circle one)

2. Check for leaks around pressure cap; note findings. _____

3. Inspect the radiator for signs of leaks; note findings. _____

4. Inspect hoses for leaks, cracks, or bulges; note findings. _____

5. Inspect radiator fan for damage; note findings. _____

6. Inspect for coolant leaks around the thermostat and hoses; note findings. _____
_____

Start the engine and set the heater controls to high heat output.

7. Does the discharge air temperature increase after the engine has run for several minutes?

   Yes _____   No _____

8. Is there any steam from the vents or is there an odor of coolant in the discharge air?

   Yes _____   No _____

Shut-off the engine, and raise and support the vehicle to check the underside of the engine.

9. Inspect for signs of coolant loss from the water pump, lower radiator hose, and lower section of the radiator; note findings. _____

10. Summarize the results of your inspection. _____
    _____

11. Based on your findings, what actions are necessary? _____
    _____

INSTRUCTOR VERIFICATION:

Lab Worksheet 26-2

Name _____ Date _____ Instructor _____

Year _____ Make _____ Model _____

Pressure-Test the Cooling System

1. Determine the correct coolant for the vehicle using the owner's manual or manufacturer's service information.

    Recommended coolant type _____

2. Locate and record the cooling system operating pressure. _____

3. With the engine cool, carefully remove the radiator cap. Inspect and record the coolant level and appearance.
_____

4. Using the appropriate adapters, install the cooling system tester onto the radiator or pressure tank. Increase pressure to the limit shown on the radiator cap.

    Instructor's check _____

5. With the system under pressure, look for signs of coolant loss from the radiator, hoses, water pump, and other locations. Record your findings. _____
_____

6. Once you have inspected the system, slowly relieve the pressure and remove the pressure tester. Clean any spilled coolant and clean the pressure tester and adapters.

    Instructor's check _____

INSTRUCTOR VERIFICATION:

Lab Worksheet 26-3

Name _____ Date _____ Instructor _____

Determine Correct Coolants

Using a selection of vehicles, determine the correct coolant for each. To determine the correct coolant, you may need to check the owner's manual, service information, and the Internet.

1. Year _____ Make _____ Model _____

   Engine _____

   Recommended coolant _____

   Cooling system capacity _____

2. Year _____ Make _____ Model _____

   Engine _____

   Recommended coolant _____

   Cooling system capacity _____

3. Year _____ Make _____ Model _____

   Engine _____

   Recommended coolant _____

   Cooling system capacity _____

INSTRUCTOR VERIFICATION:

Lab Worksheet 26-4

Name _____ Date _____ Instructor _____

Year _____ Make _____ Model _____

Engine _____

Inspect Thermostat and Determine Engine Operating Temperature

1. Using the service information, determine thermostat opening and fully open temperatures and normal engine operating temperature.

    Thermostat begins to open at _____ °F

    Thermostat fully open at _____ °F

    Normal operating temperature _____ °F

2. Connect a scan tool to the data link connector and navigate to the engine data menu. Locate the engine coolant temperature (ETC) data. Note starting temperature.

    Starting temp. _____ °F(C)

3. Start the engine and allow it to run until thermostat opening temperature is reached. Next, carefully measure the temperature of the upper and lower radiator hoses with an infrared thermometer or temperature probe for a DMM. Note the temperatures and compare them to what is displayed on the scan tool.

    Upper hose temp. _____ °F

    Lower hose temp. _____ °F

    ECT sensor temp. _____ °F

4. Based on the temperatures, is the thermostat working properly?

    Yes _____   No _____

5. If the temperatures are too low, what does this indicate? _____
   _____
   _____

Chapter 26 Heating and Air Conditioning

6. If the ECT sensor temperature keeps increasing beyond thermostat opening temperature but the temperature of the radiator hoses is much cooler, what can this indicate?

   _____
   _____
   _____

7. If the thermostat is working properly, continue to let the engine run until the cooling fan turns on. Note the temperature displayed by the ECT sensor. If the fan is belt driven, record the highest temperature indicated by the ECT sensor.

   Coolant temp. when fan turn on _____

   Maximum coolant temp. (belt-driven fan) _____

8. Let the engine run until the cooling fan turns off. Note the temperature.

   Coolant temp. when fan turns off _____

   Minimum temp. (belt-driven fan) _____

9. Based on your inspection, what is the condition of the thermostat and cooling system operation?

   _____
   _____
   _____
   _____

INSTRUCTOR VERIFICATION:

Lab Worksheet 26-5

Name _____ Date _____ Instructor _____

Year _____ Make _____ Model _____

AC Performance Test

1. Perform a visual inspection of the AC system. Note any faults or concerns with the following:
   a. AC compressor drive belt   _____ OK   _____ Not OK
   b. AC compressor   _____ OK   _____ Not OK
   c. AC hoses   _____ OK   _____ Not OK
   d. Cooling fans   _____ OK   _____ Not OK

2. Close all but the centre vent on the dash and place a thermometer into the open dash vent. Note the temperature inside the vehicle before the AC is turned on.

   Temperature _____ °F

3. Start the engine, set the air to discharge from the dash vents, place the AC on MAX cooling, and set the blower on high speed. Run the engine at 2,000 rpm and note the temperature of the air from the vent. Run the system for a couple of minutes until temperature remains steady.

   Temperature _____ °F

4. Shut the AC off and then shut the engine off. Using the service information, locate the AC performance test temperature chart. Using the chart, determine the condition of the AC system and note your results.
   _____
   _____
   _____
   _____

INSTRUCTOR VERIFICATION:

CHAPTER 27: Vehicle Maintenance

Review Questions

1. _____ is the act of keeping something in a state of good operating condition.

2. Maintenance is performed to prolong the life of something, and a repair is made to correct a _____.

3. Using the list provided in the textbook, make a list of maintenance items you already have performed on your own vehicle or someone else's vehicle. _____
   _____
   _____
   _____
   _____
   _____

4. Periodic inspection and service to keep the vehicle operating in good condition provides _____ for the technician and the shop.

5. Describe how to use maintenance as a way to build a relationship with your customers.
   _____
   _____
   _____
   _____

6. Failure to perform routine maintenance can cause serious _____ to the vehicle.

7. Maintenance information is usually located in the vehicle's _____ _____.

8. Describe the differences between what may be considered normal service and severe service. _____
   _____
   _____

Chapter 27 Vehicle Maintenance

9. Once a periodic maintenance service has been performed, it is important to _____ the maintenance reminder or timer.

10. Why is it important to review the customer's service history before recommending some services be performed? _____

11. Why should a check of TSBs be performed as part of determining what maintenance is necessary? _____

12. While tire pressure is being checked: Technician A says higher-than-specified tire pressure may indicate the tires have been recently driven on. Technician B says if the tire pressure is above the specified pressure, the pressure should be reduced. Who is correct?
    a. Technician A
    b. Technician B
    c. Both A and B
    d. Neither A nor B

13. Describe how to quickly check shock absorber condition. _____

14. Why must you use caution when replacing wiper blades? _____

15. List what lights should be checked as part of the maintenance inspection. _____

16. List five fluids that should be checked during a maintenance inspection.
    a. _____
    b. _____
    c. _____
    d. _____
    e. _____

Chapter 27 Vehicle Maintenance   501

17. Technician A says all automatic transmissions have a dipstick for checking fluid level. Technician B says most transmissions can use any type of automatic transmission fluid. Who is correct?
    a. Technician A
    b. Technician B
    c. Both A and B
    d. Neither A nor B

18. Explain why exhaust leaks can be dangerous. _____
    _____
    _____

19. Before moving any vehicle for service, always check brake pedal _____ and _____ before you begin to move the vehicle.

20. Explain viscosity as it applies to engine oil. _____
    _____
    _____
    _____

21. What does multiviscosity oil mean? _____
    _____
    _____

22. Which of the following engine oil ratings are important to understand?
    a. SAE and API
    b. ILSAC and ACEA
    c. Manufacturer-specific oil ratings
    d. All of the above

23. Which of the following are problems that may result from using the incorrect engine oil?
    a. Internal engine sludge
    b. Variable valve timing faults
    c. Engine failure
    d. All of the above

24. Before beginning an oil change, take the time to check and document the engine oil _____ and _____.

25. After removing the old oil filter, clean and _____ the filter gasket surface on the engine.

© 2014 Cengage Learning. All Rights Reserved. May not be scanned, copied or duplicated, or posted to a publicly accessible web site, in whole or in part.

26. Most modern gasoline engines hold approximately how much engine oil?
    a. Two to four quarts
    b. Four to seven quarts
    c. One to two gallons
    d. Six to ten quarts

27. Explain three ways of changing the automatic transmission fluid. _____
    _____
    _____
    _____

28. Since brake fluid is hygroscopic, the accumulation of _____ in the system leads to rust, corrosion, and copper ion contamination.

29. Limited-slip differentials often require a specific type of _____.

30. Technician A says green coolant is universal and can be used in any type of vehicle. Technician B says color has little to do with choosing the correct coolant. Who is correct?
    a. Technician A
    b. Technician B
    c. Both A and B
    d. Neither A nor B

31. Which of the following statements about coolant is correct?
    a. Some coolants contain silicates.
    b. Some coolants have no silicates.
    c. Silicates are added to coolant to prevent corrosion.
    d. None of the above.

32. New coolant should be mixed with _____ water.

33. Describe three methods of replacing engine coolant. _____
    _____
    _____
    _____

34. Explain why it is important to make sure that all air has been removed from the cooling system. _____
_____
_____

35. Some newer-diesel powered vehicles have diesel exhaust fluid _____, which requires periodic refilling.

36. DEF is composed of 33 percent _____ and 67 percent pure _____.

37. Explain the purpose of the engine air filter. _____
_____
_____

38. Technician A says the engine air filter shown in Figure 27-1 should be replaced. Technician B says the filter is probably not restricted enough to cause an engine performance problem. Who is correct?
    a. Technician A
    b. Technician B
    c. Both A and B
    d. Neither A nor B

Figure 27-1

39. Technician A says a dirty air filter can be cleaned by blowing compressed air through it. Technician B says a dirty air filter can be cleaned by shaking the dirt and debris from the filter. Who is correct?
    a. Technician A
    b. Technician B
    c. Both A and B
    d. Neither A nor B

40. _____ _____ help filter out pollen, dirt, and odors from the air entering the passenger compartment.

41. Explain the steps of performing a chassis lubrication. _____
_____
_____
_____

Activities

1. Organize the following items into two categories: those that are typically replaced as maintenance and those replaced as part of a repair.

Water pump	Timing belt
Generator	Fuel pump
Air filter	Wiper blades
Brake pads	Wheel bearing
Spark plugs	Ignition coil
Wheel speed sensor	Tie rod end
Cabin air filter	Fuel filter
Accessory drive belt	Power steering pump

2. Organize the following fluids into two categories: applications that often require specific fluid types and qualities and applications that do not.

Brake fluid	Automatic transmission fluid
Engine oil	Power steering fluid
Windshield washer fluid	Differential lubricant
Antifreeze	Hydraulic clutch fluid

   a. How can a technician determine if specific fluids are required for an application?
   _____
   _____
   _____

   b. Explain why it is important to use the correct fluid as specified by the vehicle manufacturer. _____
   _____
   _____

INSTRUCTOR VERIFICATION:

Chapter 27 Vehicle Maintenance   507

Lab Worksheet 27-1

Name _____ Date _____ Instructor _____

Year _____ Make _____ Model _____

Engine _____   AT   MT (circle one)

Maintenance Schedules

Locate the manufacturer's maintenance schedules for this vehicle.

1. How does the manufacturer define severe service? _____
   _____
   _____

2. Based on your experiences, how likely do you think it is that the vehicle is operated under severe conditions? Explain your answer. _____
   _____

3. How does the severe schedule differ from the normal service schedule? _____
   _____
   _____

4. Is the vehicle equipped with a towing package or trailer hitch?

   Yes _____   No _____

   If yes, how can pulling a trailer affect the maintenance requirements for the vehicle? _____
   _____
   _____

INSTRUCTOR VERIFICATION: _____

© 2014 Cengage Learning. All Rights Reserved. May not be scanned, copied or duplicated, or posted to a publicly accessible web site, in whole or in part.

… Chapter 27 Vehicle Maintenance

Lab Worksheet 27-2

Name _____ Date _____ Instructor _____

Year _____ Make _____ Model _____

Factory tire size _____ Tire pressure spec _____

Check Tire Pressure

1. Installed tire size

    RF _____   RR _____

    LF _____   LR _____

2. Tire pressure readings

    RF _____   RR _____

    LF _____   LR _____

3. Spare tire pressure spec _____   Spare tire pressure _____

4. What are the three side effects of low tire pressure?
    a. _____
    b. _____
    c. _____

5. How often should tire pressure be checked? _____
   _____
   _____

INSTRUCTOR VERIFICATION:

Lab Worksheet 27-3

Name _____ Date _____ Instructor _____

Year _____ Make _____ Model _____

Engine _____   AT   MT (circle one)

Fluid Inspection

1. Locate the manufacturer's recommended fluid types for the following:

    Engine oil   _____

    Transmission   _____

    Brake fluid   _____

    Power steering   _____

    Antifreeze   _____

    Differential   _____

2. Check each of the following fluids; note the fluid level and color.

    Engine oil   _____

    Transmission   _____

    Brake fluid   _____

    Power steering   _____

    Antifreeze   _____

    Differential   _____

3. Based on your inspection, what actions should be performed? _____
_____
_____
_____

INSTRUCTOR VERIFICATION: